MIRAGES
OF
DEVELOPMENT

MIRAGES OF DEVELOPMENT

Science and Technology for the Third Worlds

Jean-Jacques Salomon
André Lebeau

Lynne Rienner Publishers • Boulder & London

Published in the United States of America in 1993 by
Lynne Rienner Publishers, Inc.
1800 30th Street, Boulder, Colorado 80301

and in the United Kingdom by
Lynne Rienner Publishers, Inc.
3 Henrietta Street, Covent Garden, London WC2E 8LU

© 1993 by Lynne Rienner Publishers, Inc. All rights reserved

First published as *L'écrivain public et l'ordinateur*
(Paris: Hachette 1988; © in 1988 by Hachette)

Library of Congress Cataloging-in-Publication Data
Salomon, Jean-Jacques.
   [L'écrivain public et l'ordinateur. English]
   Mirages of development : science and technology for the Third
Worlds / by Jean-Jacques Salomon and André Lebeau.
   Includes bibliographical references and index.
   ISBN 1-55587-368-5 (alk. paper)
   1. Technological innovations—Economic aspects—Developing
countries. 2. Technology transfer—Developing countries.
3. Appropriate technology—Developing countries. I. Lebeau, André.
II. Title.
HC59.72.T4S2413 1993
338.9'26'091724—dc20   92-40754
                                                                        CIP

British Cataloguing-in-Publication Data
A Cataloguing-in-Publication record for this book
is available from the British Library.

Printed and bound in the United States of America

The paper used in this publication meets the requirements
of the American National Standard for Permanence of
Paper for Printed Library Materials Z39.48-1984.

Contents

Preface   vii

Introduction   1

PART 1 THE SETBACKS TO DEVELOPMENT

1 No Shortcut to Development After All   9
    The ups and downs of development
    "I want a fig..."
    Technology in society

2 The Many Third Worlds   23
    No blueprint for development
    The diversity of circumstances
    The noneconomic factors

3 A Basic Discontinuity   33
    From Crécy to Hiroshima
    The high-tech empire
    The experience of Latin America
    Problems with development—or with rationality?

4 The Contemporary Technical System   49
    The idea of a technical system
    From one radical shift to another
    The symbiotic relationship between science and technology
    Increasing complexity
    The human brain "off-line"

5 The Science of the Poor   63
    When universal is not universal
    Investment in research
    The magnet of pure science
    The bird with clipped wings
    The brain drain

PART 2 THE INFORMATION REVOLUTION

6 The Looking-Glass Race — 85
   The economics of the intangible
   Mastery of use and mastery of production
   The computing capacity of the Third Worlds

7 The Machines from the North — 99
   The specter returns
   Man cannot live by signs alone
   The medium is not the message
   Between nightmares and utopian dreams
   Experience and theory
   What if the robots . . . ?

8 The Cathedrals in the Desert — 115
   As a thundercloud brings the storm . . .
   Technology transfer and transport
   Weapons versus development

9 The Newly Industrialized Countries — 131
   The crucial difference
   Three giants and four dragons
   The cost of interventionism

10 History's Revenge — 147
   Unequal growth
   The hard and the soft
   Walking on two legs
   The three strategies
   Mote and beam

Appendix: — 161

Is the Third World Headed for Perpetual Dependency? — 163
   Christian Comeliau

The Advantages of Being a Late-Comer to What? — 169
   Amilcar Herrera

Science, Technology and Development — 185
   Jean-Jacques Salomon & André Lebeau

Notes — 201
Index — 211
About the Book and the Authors — 221

Preface

The main title of this book when it was originally published in French was *L'écrivain public et l'ordinateur* (*The Scribe and the Computer*), because we wished to emphasize that, while Third World countries may sometimes benefit from the most advanced technologies—such as the computer of our title—they should not abandon traditional, more labor-intensive methods (the public scribe for example), which still have much to contribute. All too often, experts in both the North and the South assume that the key to development lies in the latest science and technology, despite much evidence to the contrary—hence, our original subtitle, *Mirages of Development.*

Some other minor changes have been made in the English-language edition to update where necessary and possible (some data are now almost a decade old, but unfortunately nothing more recent is available). Clearly, the world has seen enormous changes since the book was completed in 1988, but we have resisted the temptation to indulge in major rewriting. There is one substantial difference from the French edition: We have replaced our original appendix, which sought to illustrate the essential nature of the computer revolution for a French-speaking audience in developing countries, with extracts from the debate on our book published in the journal *Social Science Information.* These extracts include the two most substantial critical reviews, by Christian Comeliau and Amilcar Herrera, and our reply. We hope that the publication of the book in English will now stimulate further discussion.

Finally, we should like to thank Ann Johnston for her careful and readable translation and the French Ministry of Culture for a substantial financial contribution to the costs of translation.

Jean-Jacques Salomon
André Lebeau

Introduction

In a stall in one of the squares of Aix-en-Provence in the south of France, sitting like a post office clerk at his window, the scribe waits for his customers. He has an ancient black metal typewriter, a far cry from modern electric machines in their brightly colored plastic cases, with memories and liquid crystal displays, equipped with such a vast range of functions, from calculation to drawing, that even the best secretaries can never hope to make full use of them. The scribe's methods are old-fashioned: He makes copies using carbon paper, which leaves smudges on the paper and ink on the fingers. But this does not really matter. He can read, write, and above all hear, and he provides an invaluable service.

This stall does not appear on any of the postcards showing the squares of Aix-en-Provence around 1900, in the 1930s or 1940s, when France was still largely rural, when the telephone, the radio, the gramophone, and the typewriter were luxuries, and all primary school pupils learned to form their letters, chanting them together aloud and writing with dip pens. In the 1980s the stall is there, set up in the square, unprotected from the elements, the scribe working on market days, just as useful today as personal computers, video recorders, compact discs, and videos. And let no one suppose that the scribe's customers are all immigrants from North Africa, Portugal, Turkey, or Yugoslavia, either illiterate or with a poor command of French. Among those who ask his help in drafting an application for a job, a legal document, or a love letter are native French citizens, born in the mountains of Provence or Auvergne or elsewhere. Some of them never learned to write, others have forgotten how, while others lack the courage to do so or else need something typed—"new illiterates" as others are the "new poor."

Astonishing progress in the techniques of information and communication can take place alongside persistent and sometimes growing pockets of illiteracy, as recent surveys have shown in both Europe and the United States. In the age of increasingly powerful silicon chips, telecommunications via satellite, and ever more user-friendly computers capable of ever more sophisticated tasks, modernity can coexist—even in industrialized countries—with practices linked to traditional and sometimes archaic lifestyles. This dualism was there already before the Information Revolution, and it continues even in the so-called advanced countries, in spite of all the changes that have occurred.

One may even wonder whether the increasing pace of technical change is not exacerbating this dualism. Not everyone who can read a printed page has to be able to use a computer; many, on the contrary, avoid doing so, and their resistance is in no way related to the refusal of schoolteachers in the past to allow ballpoints instead of fountain pens. Underlying the new tool are not simply new ways of writing but a new language and a new culture. Far from eliminating, between one generation and the next, the obvious divide between those who quickly learn to make the most of the latest techniques and those who do not, cannot, or do not want to make use of the new tools and the new knowledge, the galaxy of Edison, Marconi, and von Neumann promises instead to widen the gap.

This is all the more the case in the developing countries. Not only are different sectors of their economies changing at different rates—modern production and distribution methods coexist with preindustrial, if not prehistoric, ones—but in addition Western ways of thinking exist alongside instinctive reactions, attitudes, and institutions that Western thought has constantly rejected.

No one has discussed this dualism more ably than Albert O. Hirschman, one of the pioneers of the economics of development, who has constantly tested his theories against his experience in the field, especially in Latin America. "It is often said that the underdeveloped but developing countries are apt to pass from the mule to the airplane in one generation. But a closer look at most of these countries reveals that they are, and appear to remain for a long time, in a situation where *both the airplane and mule* fulfill essential economic functions."[1] This dualism, which involves attitudes as well as working methods and commercial behavior, is the source of tensions and impatience, yet it is unavoidable and, in some respects, it can in fact have its advantages. If, for example, these countries want to diversify their economies, they would be wise not to set up new plants to make existing products for which the returns will be meager, but instead to invest their scarce capital in industries making new products for which productivity is higher.

It is precisely in those industries and technologies which demand the greatest efficiency (in particular, those requiring careful maintenance) that the underdeveloped countries have the best chances of success. Hirschman first made this point in 1958, and the example of transport consistently proves it: In these countries, the airlines tend to run well, the railroads provide a mediocre service, and the road network is in a shocking state of disrepair. Consequently, the most modern industries, which cannot tolerate negligence in their operations or falling production standards, can ensure "a comparative advantage in jobs that must be done well if they are to be done at all."[2] These forms of capitalistic production not only encourage higher productivity, but also speed up the growth of some industries and make larger areas of the economy more competitive.

On the other hand, as soon as one turns from production to management tasks, low operational standards rarely lead to catastrophe—train derailments or air crashes. The level of care and discipline required is more vague, the margin of tolerance of inefficiency is greater, and progress inevitably remains marginal. Whether this involves problems of organization, of financial management, of relations with staff and clients in private firms, or of public administration in general, the possibilities of success are all the smaller because nobody is very sure about what criteria they should be measured against.

> National character and history are usually appealed to in explaining the malfunctioning of the political and administrative processes which stands so often in sharp contrast to real achievements in industrial and agricultural production. Seldom is it realized that these processes are intrinsically harder to master than production jobs. On the contrary, amazement is expressed that a country pretends to set up modern industries when basic problems of public administration have not yet been solved. But this is only one of the many instances where what seems a cart-before-the-horse sequence turns out to be the efficient one in underdeveloped countries.[3]

Experience shows that choosing "to put the cart before the horse" has indeed led to successes in both the public and the private sectors; it partly explains the increasing strength of the newly industrialized countries. Yet it is impossible to generalize: For one thing, there have also been some outstanding failures; for another, this speeding up of the process of industrialization, whatever the successes, has merely exacerbated the inequalities of the dual society in every instance. As Hirschman himself admitted, more than a quarter century after his first (and convincing) studies, this choice is frequently offset by a very high price: "The major disappointments of the past two decades over Third World developments have occurred in the political realm. While the economic growth record has been far from fair to excellent, at least in terms of aggregate expansion, the political record must be called from barely tolerable to disastrous."[4]

This assessment also explains why, during the same period, the paradigm of development economics has changed totally. In the past, the emphasis was on growth rates, industrialization, and international aid; now the stress is on income distribution, employment creation, and self-sufficiency. In shifting from one paradigm to the other, we have moved from a concern with economics and production defined in terms of the double-entry bookkeeping of development plans, to an approach that takes greater account of anthropology, social psychology, and even the wisdom of nations. All in all, the experts have learned a great deal about the limits of their models when applied to the situations of the Third World, as Hirschman notes with a touch of irony: "In that eminently 'exciting' era, development economics did much better than the object of its study, the economic development of the poorer regions of the world, located primarily in Asia, Latin America, and Africa."[5]

Does this mean that people have really become aware of the link that can exist between choosing to put the cart before the horse and the political and social disasters that have occurred, *in spite of* the satisfactory results in terms of growth achieved? This seems far from the case, given the literature that has poured out in recent years describing all the benefits the spread of information technologies must bring to the developing countries. The Information Revolution is presented in these publications as the ultimate weapon for solving all the problems and for catching up.

Among this literature that veers between utopia and myth, there is no more striking example than Jean-Jacques Servan-Schreiber's *The Global Challenge*. One may read there, for example, that "the revolution of the microprocessor and of telecommunications has provided the means to speed up, in a hitherto inconceivable fashion, a process of development that gives hope of achieving equality, not any longer in 150 years, but perhaps in a single generation."[6] Where will the money come from to finance this manna of computers and software designed specifically to meet the needs of the poorest countries? From oil wealth—apparently a philanthropic alliance of Gulf sheikhs and Japanese industrialists will take up this challenge. Thus the Third World will soon be flooded with communications equipment, and everything will become possible, because the machine will truly become the message:

> There can be no question of proclaiming that reading and writing will no longer be taught. Rather it is a case of observing that any human being who can hear and speak will be able to communicate with a microcomputer, and therefore will be able to take part in what is going on, relying only on his ability to think, which he shares with the rest of the human race.[7]

Unfortunately for modern prophets, the pace of change—less rapid after all than they had promised—is enough to reveal the lack of substance of their predictions in their own lifetimes. As we shall see, it is not a case of underestimating the contributions the information technologies could make to the development of the Third World. But even if they could do a great deal, they cannot do everything and anything. They are not the key to catching up, and it is wrong to proclaim the contrary *urbi et orbi*. Although they may be revolutionary, they cannot alone trigger or maintain the social transformations that shape the possibility of development.

Beyond a certain minimum satisfaction of basic needs, capital accumulation by itself is never a guarantee of growth, and if science and technology can stimulate the development process, the political and social context must be favorable and therefore must be ready for it. Indeed, it is the way the society is organized that determines the area in which scientific knowledge and technical innovations generate growth—and not vice versa. The "computer fairy" can certainly accomplish a lot, but scientific research and technical innovation produce rapid results only where the organization, institutions, and attitudes have previously removed most of the obstacles characteristic of traditional economies and societies. Everywhere else, what happens to these obstacles will be what determines the pace of change—and there is no magic wand to remove them at a stroke.

A word of warning is in order here. Most of the discussion in this book refers to the nation-state. There is little alternative, because most of the quantitative data (especially those provided by the international organizations) are produced on this basis. Yet it is worth stressing the limits and the paradox this involves. In the Third World, the nation-state—a relic of the colonial period—does not mean the same thing as it does in the industrialized countries. The term obscures, sometimes at the cost of bloody conflicts, much older ethnic and tribal divisions and simultaneously prevents the creation of larger units. "Nation-state" is also inadequate to describe properly the social and political complexities of the countries of the Third World, but at the same time it explains why it may be so difficult to bring about cooperation among several of them within a region, even though in the area of science and technology this would be one way of overcoming the shortages of money and skilled labor that hinder the development efforts of all the younger nations.

Even in the case of the industrialized countries, it is obvious that the combination of the trends in the technical-industrial system and international trade leads to the creation of vast structures for research, production, and distribution that are less and less compatible with the strict observance of the principles of sovereignty that underpin the nation-state. The smaller the country, the bigger the fundamental problem: the choice between whether to accept the loss of power to organizations beyond the reach of national sovereignty or instead to risk economic decline. Judged by the number of agreements on scientific and technical cooperation that have been signed—not to mention the afterthoughts, the touchiness, and the difficulties these agreements have encountered—it is clear that European countries have not managed to avoid the contradictions inherent in this choice. In fact, only the continent-sized countries have been less affected by this problem so far.

This brings us to the paradox: For all that the growth of transnational organizations, both public and private, gives concrete expression to this globalization of the technical-industrial system, the will of the nation-state remains the most decisive stimulant in the attempts by certain countries to compete with the most industrialized nations. What is true of some European countries is even more so of the continent-sized countries of the Third World: Brazil, China, and India. Nevertheless, the statistical data on the scientific and technological potential of each country taken on its own provide an extremely inadequate measure of the pressures and hindrances each must deal with when trying to make the most of this potential.

If the transnational megastructure—from multinational firms to international cooperation agreements—is indeed becoming the characteristic feature of the technical-industrial system, then it is obviously a makeshift solution to analyze current developments in terms of the nation-state, which this phenomenon is weakening or destroying. The choice of these terms of reference does not mean we consider that maintaining the integrity of the nation-state is an aim in itself; national boundaries are merely a constraint to which technical and economic progress is less and less well adapted. This does in any case bring out the narrow room for maneuver available to the developing countries.

This book sets out to be a return to common sense, an approach that has not always guided the debate. While we stress the opportunities created by the information technologies, we try to show that they also involve limits, threats, and traps—for the industrialized countries, let alone for those of the Third World. The book is based on our experience with these problems in certain developing countries and on our involvement in an international program concerned with the impact of the new technologies on developing countries.[8] Its message in brief is to point out that if the scribe and the computer still exist side by side in our countries, then they are all the more likely to do so for even longer in the Third World. *The shortcut to development is never science and technology in themselves, but is development itself.*

Part 1

The Setbacks to Development

> Research carried on in the rich countries on the problems of underdeveloped countries . . . tends to become "diplomatic," forbearing and generally overoptimistic.
> —*Gunnar Myrdal*

1 No Shortcut to Development After All

Whenever the facts do not fit the theory, it is normal to look for the error in the theory rather than in the facts. This is indeed the essence of all scientific method: not to read the world in terms of theoretical preconceptions but to observe what is actually happening. Facts are stubborn, as Lenin said; yet the people he inspired have been rather too inclined to dismiss any facts that did not fit their theories. Because facts are obstinate, the only way to challenge them is to produce others, created out of thin air if need be, which is how ideology can conjure up fantasies.

The idea that the proletariat in the industrialized countries was inescapably condemned to increasing poverty was one of these fantasies. In France, for example, during the period of rapid growth in the 1950s and 1960s, the more the statistics indicated that everyone was getting richer, the more the Marxist intellectuals, trade unionists, and political activists thundered about the worsening situation of the workers. Although the purchasing power of manual workers' net wages was clearly rising (by 50 percent between 1951 and 1965), they continued to say the exact opposite.[1]

Similarly, there was alleged to be a close, and indeed automatic, link between the industrialized countries becoming richer and the developing countries becoming poorer. This process, too, was seen as inevitable and as the direct result of the neocolonial economic imperialism of the rich nations, just as the pauperization of the proletariat was supposed to be the conscious and deliberate action of bourgeois capitalists. In both cases, if the situation deteriorates, it must always be the consequence of malicious intentions: For workers in the industrialized countries, as for the proletarian nations, increasing poverty is deliberately built into the capitalist system, just like a worm in an apple. What then does one do if there are cases where grinding poverty has diminished or if the setbacks to development today are not in the least like the burdens of colonial exploitation? In that case, the statistics must be wrong.

The Ups and Downs of Development

Anyone who travels in the Third World or reads the reports of the international organizations is aware that remarkable progress has been made, yet the results differ so widely that it is unwise to use them to make general predictions about the future. There are still too many places where food production lags behind population growth, even though many more countries than before have managed to increase their level of self-sufficiency thanks to the spread of high-yield crops and of agricultural advisory services.

Neither India nor China, for example, is in the least what it was less than a quarter century ago, when both suffered from frequent famines, epidemics, and natural disasters. The introduction of new varieties of grain and the improvement in farming methods generally in India, Pakistan, and Indonesia have speeded up their transformation into producers of surpluses. In China, after the upheavals of the Cultural Revolution, the stimulus of a moderate dose of private enterprise and competition helped a large section of the rural population to improve its position by at least 50 percent between 1980 and 1984.

These spectacular results derived above all from better irrigation (more efficient pipes and channels, many more pumps installed on wells and riverbanks) and the huge increase in the number of high-yield varieties of wheat and rice. Between 1950 and 1980, the irrigated area in India rose from 50 to 100 million acres and in China from 40 to 80 million. Countries as different as Burma (now Myanmar), South Korea, Pakistan, and the Philippines doubled the area under irrigation. These results also depended upon the quality of the people running the agricultural sector and whether they encouraged the spread of new techniques arising out of agricultural research. Another contribution came from the implementation of a more efficient grain policy based on better management of stocks and some intervention as regards prices and markets (in China this meant in fact a limited reintroduction of free-market mechanisms).

These successes in Asia are in sharp contrast to the equally spectacular failures in Africa. Nevertheless, thirty years ago it was about the desperate food shortages in overcrowded Asia that the experts were raising the alarm, whereas the situation in Africa, where population growth was slower, did not worry them as much. Both the assessment of the current situation and the forecasts made by Edouard Saouma, head of the United Nations Food and Agriculture Organization (FAO), in a report published in 1986 are catastrophic: Although they were virtually self-sufficient 25 years ago, most African countries cannot now feed themselves; unless there is a radical shift in the policies of the countries concerned, along with massive aid from the richer nations, the situation can only get worse by the end of the century. Since 1961, per capita food production has fallen by almost 20 percent and the population explosion is frightening: over 3 percent per annum on average.

Why are there empty stomachs in one area and full storehouses in another? Virtually all the experts now agree in seeking the main reasons in the social and cultural background of the countries and in policies detrimental to agriculture. This does not mean we should discount the physical constraints imposed by climate and geography or the consequences of fluctuations in the prices of primary products. Nevertheless, drought and the deteriorating terms of trade are useful alibis to disguise problems that have nothing to do with the state of the heavens or of the commodity markets. The *internal* explanations for the setbacks, which are beginning to be recognized by the countries themselves, are far more weighty than any others: the contempt for food crops, exhaustion of good soils, destruction of forests (10 million acres per year!), lack of trained people to educate and advise farmers, rural institutions stifled by overgrown bureaucracies, the attraction of the cities and of work in service industries, too many officials, political instability, governments without political legitimacy, and so on.[2]

Nonetheless, taken overall, the developing countries have made undeniable progress. Between 1960 and 1984, the average annual increase in gross domestic product (GDP) per capita was about 2.8 percent (not including China and some oil-exporting countries). As for the oil-importing countries, the highest growth has been in Latin America and the Far East, with rates well above the overall average of 3.4 percent for all the developing countries taken together. These rates have been 50 percent higher for a quarter century than was the case for today's industrialized nations during a whole century of development: 2.7 percent on average between 1850 and 1960.[3]

Clearly there are enormous differences among countries and regions. The lowest rates of growth (2.9 to 1.8 percent) are found in the poorest countries of Africa and Asia. The rise in oil prices and the recession in the 1980s made the economic situation of some countries even worse than before. In Africa and Latin America, at least a decade of increase in per capita income seems to have been lost for this reason. In Asia, on the other hand, many countries have achieved impressive rates of growth (8.6 percent annual average increase throughout the period 1960–1982).

From the disasters in Ethiopia and the Sahel region, to stagnation if not decline in Ghana or Zaire, to the encouraging results in countries like India, Brazil, Indonesia, Malaysia, or, more recently, China, the disparities in the growth rates of per capita incomes reflect the range of situations as regards population growth, resources, and above all the choice of economic and political strategies. Nevertheless, twenty-seven countries more than doubled their per capita GDP between 1960 and 1982: Ten are oil exporters (Algeria, Ecuador, Egypt, Indonesia, Iraq, Libya, Mexico, Nigeria, Saudi Arabia, and Syria), and five are classed as newly industrialized countries, or NICs (Brazil, Hong Kong, Singapore, South Korea, and Taiwan).

Economic growth has (far) outstripped the forecasts generally proposed twenty-five years ago. In spite of the recession, at the beginning of the 1980s the developing countries were producing six times their 1950 output of goods and services, and their manufacturing output recently reached a figure eight times that of 1950. These countries' manufactured products increased their share of the Organization for Economic Cooperation and Development (OECD) market from 7.1 percent in 1955 to 17.8 percent in 1981. In human terms, the average life expectancy in developing countries rose from 42 to 59 years between 1950 and 1980 (from 40 to 56 years if China is excluded), while infant mortality among children under the age of four fell from 28 to 12 percent (including China).

These improvements in the Third World as a whole cannot be denied, even if the enthusiasm of the applause should vary to take account of the considerable differences among countries and—above all within each country—between those social groups that have genuinely benefited and those that have been left behind. Herein lies the essential difference between the industrialized and developing countries: In the 1950s and 1960s, growth in the industrialized nations substantially reduced the incidence of poverty, whereas in the developing countries growth has increased it. Policies aimed at development, in those instances where they were followed resolutely, have been hindered by the twin problems of a soaring birthrate and a highly skewed income distribution.

The true outcome is therefore not clear-cut, and the successes remain tiny in comparison with the scale of problems still to be solved. Full storehouses on their own do not fill empty bellies. Nobody any longer dies as a result of famine in India or China, and that is a major advance; but the general (though unquestionable) improvement does not prevent malnutrition linked to poverty from continuing to exist. The full stores are a sign that the policy of increasing agricultural production has worked, but they do not prove that the crucial challenge of development—the fact that huge sections of the population are close to destitution—has been dealt with. "Neither the growth of agricultural production nor self-sufficiency in food defined by the criterion of imports and exports alone, nor the existence of sizeable stocks is therefore a guarantee that malnutrition has been conquered."[4]

The excessive enthusiasm of most countries for development based on heavy industry, exploitation of mineral resources, and production of consumer goods has in fact made the distortions in their economies and the social disparities all the greater. It was thought that the surest way to economic "takeoff" was to focus on manufacturing, but when this eventually happened, the benefits of growth did not spread to reach most of the population. This choice of priorities, for which the industrialized countries and the international organizations are partly responsible, led to a substantial difference between the rates of growth of manufacturing and agricultural production.

The rate of increase in food production has been totally inadequate in comparison with the enormous population growth in the developing countries (over 1 billion extra mouths, of whom 600 million were in Asia, between 1965 and 1983). Even in those countries where food production grew faster than the population, the poorest people living in rural areas, who make up the majority, were not much better off. And in the cities and towns, the expansion of manufacturing and services has not generated enough new jobs to absorb the increase in the numbers of jobseekers.

Consequently, progress in economic terms did not stop the numbers of poor people from rising. In 1980 the World Bank estimated that about 800 million lived in extreme poverty, almost half of them in India, 130 million in sub-Saharan Africa, another 130 million in South Asia and the Far East, and 140 million in the Western Hemisphere, North Africa, and the Middle East. The problem of poverty in the Third World, whether seen in terms of production or of incomes, is still essentially one of rural underdevelopment, despite the continuing rise everywhere in the proportion of the population working or looking for work in the cities.

It is in the light of this state of affairs with all its contradictions—encouraging in some respects, depressing in others—that questions should be raised about the true scope for possible "shortcuts," that is, the idea that there are ways of catching up to the industrialized countries by quicker routes than the century (at least) that the latter took to industrialize.

The very notion of "catching up" is not that obvious: Is development really a race in which the laggards can hope to draw abreast of the front-runners? Putting the question in these terms does not mean that we regard underdevelopment as unavoidable; rather we are suggesting that there are other routes to development besides those followed by the most highly industrialized countries, and in particular that it is not necessary to rely heavily on the most advanced technologies. The overall progress achieved so far, and by the newly industrialized countries in particular, shows that the struggle to modernize is far from hopeless. But is "the last shall be first" true for the race between nations?

There is, of course, the example of Japan as a country that has made the transition from having a feudal society and a basically rural economy to being among the leading exporters of advanced manufactured products. But this has not happened overnight. Japanese industrial might is not a recent or even a postwar phenomenon; the Japanese had absorbed, and indeed rapidly mastered, Western scientific and technological knowledge well before Pearl Harbor. With the Meiji revolution/restoration (1868),

Japan started to industrialize only just after France, and at that time already had acquired the vision, the means, and the discipline to implement what was planned as a long-term effort.

Among the components of this effort, none was more crucial in the long run than the policy on education. The Japanese deliberately chose to import, adopt, and adapt the kinds of education and training they felt had proved the most successful for the purposes of industrialization in Europe and the United States. The University of Tokyo was established in 1877, when the first Japanese scientific societies were also created, and the Imperial Academy of Sciences was founded two years later. Japan effectively stepped into the modern world when the samurai, forced to abandon their feudal rights, took the decision virtually collectively to switch their energies to studying science and technology. They were the dominant social group, but there were not enough posts for all of them in government, and only about 10 percent of them could make their careers in public life. Because they could not compete with the other castes in the traditional areas of agriculture, handcrafts, and trade, the former warriors—who had been aware of Western technological superiority ever since Commodore Perry's visit and the opening of the ports—made themselves the pioneers and the supervisors of industrialization.

While it is true that the first graduates of the Imperial University in science and engineering were drawn from the former warrior caste, we should overlook neither the fact that the social and economic organization of pre-Meiji Japan was well suited to the process of industrialization, nor (more important) the role of the education and training policy steadfastly pursued for over a century. From this angle, many European countries did not embark on their Industrial Revolutions much ahead of Japan.

People talk about the "Japanese miracle" as if it sprang from nothing, like Athena from the head of Zeus, fully formed and not needing to mature. In fact, however, the Japanese realized early on that industrialization required coordination between the trends in technical progress and the pace of training a skilled workforce at every level of activity. They were not content simply to concentrate on training or retraining top managers: Economic growth was accompanied by a commensurate increase in the numbers of middle managers and skilled workers. To see a "Japanese miracle" is in fact to be blind to the determination—and the scale—with which Japan planned, implemented, and maintained its policies for education and training *over the long term* in order to meet the growing demands of industrialization.

"I Want a Fig . . ."

The real question raised by the strategies of looking for shortcuts is, What are the costs involved? "Leapfrogging" is possible, obviously, but experience shows that this almost always means progress for a tiny group within the country concerned. Efforts to achieve rapid industrialization, connected with the latest technologies, have led everywhere to a substantial increase in the disparities between the few who share the gains and the vast majority of the population who at best can only gather the crumbs. It is hardly surprising if this approach, through which only a small section of the urban population gets the direct benefits, comes up against the inertia of the rural masses, if not signs of resistance and outright rejection.

One of the present writers, at the time in charge of science and technology policy at OECD, remembers a meeting there just after the Yom Kippur war when European countries were fearful of oil shortages and the West was even more worried that it would not be able to "recycle" the oil revenues suddenly built up by the Arab producers. The secretary-general of OECD had an official visit from the Iranian planning minister, and the heads of various departments were invited to a meeting where they were asked to listen discreetly to the possible requests of the shah's representative. At first those around the table made cautious remarks about the macroeconomic situation, international trade, holding down or raising oil prices, the repercussions on the growth rates of the industrialized countries, and the ultimate impact of these on the developing countries.

The meeting then moved on to more serious matters when the shah's representative asked outright what the OCED could offer if Iran were to become involved in its work. "How can you help us? We have enormous resources, we have huge holdings of dollars, and we shall be earning more and more. You have the technology, the technical experts, the know-how. We are a valuable partner, of all the developing countries the one with the best chance of rapidly becoming an industrialized nation. Furthermore we are solidly on the side of the West."

We had all seen full-page advertisements in the major European and U.S. newspapers announcing that before the year 2000 Iran would be one of the five leading industrial countries in the world. Should this propaganda message be taken seriously? At the time of the celebrations at Shiraz, more magnificent than the Field of the Cloth of Gold, the shah had declared that the Pahlavis, who had already ruled Iran for twenty-five centuries, would occupy the throne for at least another millennium. Nonetheless, although the example of rapid industrialization then offered by Iran was fascinating, many cracks were already appearing in the newly restored Persian dynasty.

One might wonder about the regime and the support it enjoyed among the middle classes, all the more so because students studying abroad passed on tales of repression. There was even greater reason to wonder about the support for this high-speed industrial revolution, imposed without regard for the ordinary people's traditions and beliefs. Most of the technical experts were foreigners—not only the scientists and engineers but also the work supervisors and even the truck drivers who brought the materials needed to build the huge plants. The stallholders of the Tehran bazaar grumbled at the idea of highways that would cut across and transform the city. People were more worried than delighted at the prospect of Americanization. And the Imam Ruhollah Khomeini, the symbol of a revolution that (yet again, after Stalin and Mao) attracted the European Far Left with its promises of redemption for the West's guilt, constantly incited his people to holy uprising (all the while spreading his message thanks to an invention of Western technology, the minicassette).

"As someone responsible for science, don't you think that with new media and computers it will be possible from now on to train scientists and engineers far more quickly than you do in the West?"

Convinced that it was absurd to put the question in these terms, the man responsible for science and technology policy looked for a diplomatic reply—the least he could do since OECD was acting as host to a representative of a nonmember country: "You know, Minister, the history of science tells us that it takes time, a long time, to build up a scientific institution and train generations of researchers."

Clearly annoyed, the minister turned to the secretary-general and said: "What is the history of science doing in an organization dealing with economics? Don't you think that all of that is, to the say the least, old-fashioned? With the new technologies we shall have ten times as many research scientists in less than ten years, and our laboratories will catch up with yours!"

After all, wasn't the ancient Iranian dynasty—with its army, police, refineries, and young universities—copying the best model, the United States? What, then, was the point of history of science in the affair? Doubtless the proper reply was to say that of course the shah's representative was quite right to force the pace, or at least to think he could do so with impunity. Nevertheless, admittedly rather impertinently, the science policy chief could not stop himself from quoting Epictetus who, when one of his disciples said "I want a fig," remarked "That takes time."

The minister did not survive the upheavals of the revolution, the ayatollah's men seized power, and the ambition to make Iran one of the five leading industrial nations by the end of the century vanished. If the fundamentalists managed to defeat one of the most modern police and armed forces, was it not primarily because of the resistance the hasty industrialization generated among the Iranian people, more susceptible to the voices of prophets than to the forecasts of planners? Khomeini's supporters denounced the instruments of imperial power for the fact that they embodied the Western values of "rationalism and atheism." By destroying and purifying these instruments, the new institutions put in their place (Revolutionary Council, Islamic Committees, Revolutionary Guards) aimed to make a complete break not only with the old political setup but also with all its economic, scientific, and cultural policies contaminated by Western ideas.

Islam does not, however, have a monopoly on fundamentalism, any more than the fanatical Shiite clergy that of terror. Right or Left, have all the fundamentalists read their Jean-Jacques Rousseau? It is striking the way all of them rail against the scientific establishment. To be more precise, they attack those scientists who are apparently closely linked to the economic and political leadership (although often, in fact, this is not the case at all) and who symbolize the methods and the knowledge that leadership relies on to maintain its position. Whatever the creed, be it the Koran or *Das Kapital,* the fundamentalists always put the "intellectuals" and the "experts" on trial in the name of salvation, of either the soul or a class.

It is easy to see what drives this trial: Western rationalism embodies both a means of domination and "disillusionment with the world" in the Weberian sense. Science disillusions because it provides only instrumental answers to the questions put to it. On the other hand, its answers "work" and are the exact opposite of the Word, prayer, and faith. Max Weber maintained that adopting a rational view was linked with this disillusionment.[5] This makes a double wrong. The mastery of natural phenomena causes the disappearance or distancing of charms, magic, dreams, or faith as means of action; moreover, scientific rationality constantly proves its effectiveness and is identified with dynamism and productivity, the strengths of the industrialized countries. Consequently, science is perceived as associated with the forces of evil—the bourgeoisie, capitalism, colonialist Europe, U.S. imperialism.

In China during the Great Leap Forward and the last years of Mao's reign, just as in Iran after the advent of Khomeini, science was one of the main issues in the struggle of the "right-wing" technicians and professionals against the forces of popular revolution. The Red Guards that Mao sent into the schools and universities challenged the "paper tigers" of Western rationalism in exactly the same way as Khomeini's Revolutionary Guards. Nevertheless, although the Maoists were more concerned to maintain ideology rather than technology at the controls, and politics rather than economics, even the Great Helmsman never went so far as to do without technical experts in military research and materiel.

In the midst of the worst upheavals of the Cultural Revolution, the bridling of the teachers did not extend to removing the special immunity—or to use a term that has strong historical resonances in China, the "concession"—granted to specialists in nuclear weaponry and advanced technologies. Mao opposed the pragmatic tendencies of Deng Xiaoping, then secretary-general of the Communist Party, who is said to have made the distinctly unorthodox statement in 1962: "It doesn't matter whether the cats are black or white, just as long as they catch the mice." The cats had to be red, but not necessarily scarlet if that meant endangering the building of atomic mousetraps. Consequently, the major programs relating to nuclear and space research and computer technologies were given "extraterritorial" status, and they do not appear to have suffered unduly from the Red Guards. By contrast, agricultural production and industrial growth collapsed during the Great Leap Forward and the Cultural Revolution, as a result of campaigns conducted in the name of "mass equality."[6]

Perhaps as the tribute of vice to virtue or of the force of circumstances to ideology, the Cultural Revolution does not seem to have radically altered the policy toward certain professionals. The scientists were left to work on the major programs alone, while at the same time university teachers became the main target of attacks. Fundamentalism of whatever kind clearly sees in Western know-how reasons for fighting against the elitism and expertise of the technicians at the controls, yet it also knows how to sup with the Devil if he has knowledge that determines who the winner will be. The virulent denunciations of "atheist capitalism, Zionism, and Europe sold to U.S. imperialism" by the ayatollah's supporters did not prevent the purchase of arms from sworn enemies—via Israelis, as we now know. In the ideological battle between the two camps—right-wing experts versus left-wing ideologues in China, realist managers versus fanatical mullahs in Iran—mobilizing the masses is never enough to cause the force of circumstances to retreat before the spells cast by words.

Technology in Society

Why is it that modern science, as we have known it since the seventeenth century—since Galileo—has developed only in Europe and not in China (or India)? And why is it that until the fifteenth century the Chinese were far more efficient than the Europeans at applying their knowledge of nature to practical human needs? According to Joseph Needham, the great Cambridge scholar who has written a superb history of Chinese science and technology, the answer to both these questions lies above all in the social, intellectual, and economic structures of the different civilizations.[7]

The history of science is not alone in showing that certain societies or civilizations, at particular periods in history, are more efficient than others in their mastery of scientific knowledge and the exploitation of technical progress. In our own time, while people talk about the new technologies (computers, telecommunications, biotechnologies, new materials) as a further stage in the Industrial Revolution, it is obvious to everyone that there are wide differences in the capacities of different societies to take advantage of the possibilities opened up by the new technologies and, even more, in their ability to contribute to the conception, development, and production of new products and processes.

While social and cultural factors—from attitudes and beliefs to economic, political, and social organization—affect the role that science and technology play in a given country, the spread of new knowledge, products, and processes developed thanks to science and technology in turn transforms social structures, behavior, and attitudes. Technology itself is a *social process* among many others: It is not a question of technical development on one side and social development on the other, like two entirely different worlds or processes. Technical change and society constantly interact and alter one another. In this sense, a society is defined no less by those technologies it is capable of creating than by those it chooses to use and adapt in preference to others.[8]

Naturally, the rapid spread of a new technology does not by itself imply rapid social change. Other factors are involved, such as economic and social policies, education policies, the negotiations and agreements among interest groups, the well-established habits of daily life and of social institutions, the society's values and traditions. In the simplest terms, it can be argued that the impact of a technology on the whole society, in terms of pace and degree of penetration, depends on the interaction of four sets of factors:

Scientific and technological factors. These factors are closely linked to developments in public or private laboratories and are basically limited by the abilities of the researchers, the quality and quantity of their equipment, the flexibility of the organization, the scale of financial investment, and so on. It is, however, essential to distinguish here between truly basic research, which does not aim to produce immediately usable results, and all other forms of research ranging from the development of prototypes through applied research and development, which set out to generate useful results in the short or medium term (even if the nature and the scope of the final outcome are often hard to foresee). Technical innovation puts the seal of approval upon the market in new ideas, products, or processes. Scientific research often plays a lesser role in the success of an innovation than do ideas about management or organization, design, or marketing and advertising.

Economic and industrial factors. Discovery and invention are the province of the researcher; innovation is that of the entrepreneur—and the financier. Economic and industrial factors are thus the most influential in a society's capacity for innovation. An economy in recession is not likely to attract investments: For firms, lack of capital, plant costs that cannot be written off, and the unwillingness of investors to take risks are all factors that can prevent or delay the success of a new product or process. Clearly there are economic and fiscal policies that are more likely than others to stimulate technical innovation; industrial structures more suitable than others to take advantage of technical changes in competing on national and international markets; social circumstances and attitudes more open than others to the notion of risk taking.

Social factors. Even when an innovation seems to be economically advantageous for a firm or a country, its introduction may be postponed or ruled out because of the values and behavior of potential users. The thinking of inventors and producers does not necessarily match the ideas of consumers and users. People have to become accustomed to new technologies, so that adequate time and appropriate education are needed if innovations are to spread without stirring up too much resistance. Moreover, adopting a new technology may involve risks and costs that are often borne by only a fraction of the population (working in mines, for example), while the benefits are widely distributed (growth of energy resources). This imbalance between the costs and benefits can also work the other way, as, for example, when polluting industries create nuisances for the whole community that outweigh the local or individual advantages (for the industrialists but also for the workforce concerned to retain its employment). In the name of the collective interest, technical change obviously must be subject to public regulation (as for new drugs, for example) and the limitation of competition lest some innovations lead to greater drawbacks than advantages.

Institutional and cultural factors. Institutions, regulations, and collective beliefs exist in order for society to maintain a certain equilibrium in the distribution of risks and benefits. But innovation is by definition something that challenges habits, received ideas, and traditional values. On one hand, some regulations can inhibit the spread of a new technical system (cable television networks, for example); on the other, some technologies may upset the beliefs and values of a group or a whole society (such as the contraceptive pill). The degree of receptiveness or resistance to the introduction of new technologies depends on the nature of the technologies in question, the sociocultural environment in which they seek to establish themselves, and the timing of their introduction. Just as there are innovations that are not ripe for a particular community or society, so there are communities and societies that are not ready for a particular technical innovation—although the situation may change in the course of time.

Each of these four groups of factors should be studied carefully in any attempt to assess the impact of a new technology on society. But it is much more difficult to try to understand the way in which each interacts with the others in order to be able to analyze, let alone predict, the pace and extent of diffusion of an innovation. The task is made even more complex in that innovation is a dynamic process, and allowance should be made for the social and cultural setting in which the process takes place. In some industrialized countries, for instance, video systems were technically feasible, but efforts to launch them failed because institutional factors came into play before the economic and social factors had been properly understood and taken into account. Another example is the resistance of environmentalist groups in the United States to the construction of new nuclear power stations, even though the economic and industrial factors strongly favored the expansion of this form of energy in the 1970s.

All in all, there is no inevitability in technical change: Neither its pace nor its direction is predetermined (even if it is unwise to underestimate the strength of certain nations or industrial lobbies in imposing their factories or products), and the success of an innovation is never assured. Technology influences economics and history, but it is itself the product and expression of culture. The same innovations can therefore produce very different results depending on the structure and values of the society (or at different periods in the same society). "Technological change is often discussed as if its rate and direction were something predetermined, and as if it were something to which individuals and society could make only rather passive adaptations," Nathan Rosenberg has written. "Actually, these things are largely the outcome of a social process in which individuals and larger collectivities make choices determining the allocation of resources, and these allocations inevitably reflect the prevailing system of values."[9]

Another distinction needs to be drawn, this one between the impact of technical progress on the industrialized countries where these advances are initiated, and the penetration of traditional societies by imported technologies. The local value system in the latter case is in direct conflict with the foreign (often, indeed, alien) values conveyed by the techniques developed according to Western thought processes. Traditional societies have so little freedom to choose the technologies appropriate to their needs that their propulsion into the modern age often seems more endured than desired. Hence arise the limits and sometimes the failures of certain experiments in modernization conducted at headlong speed, without regard for the economic, social, or cultural realities of the societies into which they are being introduced. The utilization of science and technology cannot be reduced simply to inserting know-how, techniques, and methods into a social fabric that has not been prepared beforehand. We shall examine later the equivocal nature (some would say the illusion) of the notion of technology transfer. Transfers of technology require much more than the movement of a physical object from one place to another. Learning to use new technologies has a social dimension: Each new tool involves a new type of organization, a new discipline, a new way of working. The keys to using and maintaining a transplanted technical system are not to be found in the instruction manual provided by the supplier but in the education and training of the user. To think that one can skimp on this training is to run the risk of turning the keys in vain or of breaking the machine.

Technology never acts alone in the process of economic and social development, and the success of efforts to introduce a new technique in a given situation will depend mainly upon the way it is introduced and made familiar to the people concerned. This makes it all the more necessary to emphasize the enormous diversity of the developing countries. The fact that they face similar problems does not diminish the great variety of their historical backgrounds and endowments. This aspect must now be examined if we are to understand both the problems and the stakes involved as these nations confront the new technologies.

2 The Many Third Worlds

The concept of development, applied to the nonindustrialized countries, evolved after World War II and was based on three things: improvements in economic statistics; decolonization, which increased the number of newly independent nations; and the model the industrialized economies offered to those that were then called bluntly "the underdeveloped countries."

As Gunnar Myrdal has pointed out, the vocabulary of the social sciences is not neutral: To talk about "developing countries" rather than "underdeveloped countries" tends to minimize the reality of their structural differences and to put the emphasis instead on the possibility of catching up. The courteous diplomatic phraseology gives the impression that there is just a simple time lag between countries that are industrialized and those that are not—and the "right" economic policy will soon do the trick. As Myrdal says:

> The term is, of course, illogical, since, by means of a loaded terminology, it begs the question of whether a country is developing or not or whether it is foreseeable that it will develop. Moreover, it does not express the thought that is really pressing for expression: that a country is underdeveloped, that it wants to develop and that perhaps it is planning to develop.[1]

It should not be forgotten that the developing countries became members of the United Nations in the context of the cold war and the division of the world into two opposing camps. The new nations immediately became pawns in the foreign policies of both East and West, subjected to rival offers, aid and pressures from their communist or capitalist guardians. Whatever their creed, Marxist or not, these guardians agreed on at least one thing: They were equally optimistic about the possibilities of catching up.

Perhaps they were even more similar in (i.e., equally taken in by) the simplicity of the purely economic solutions proposed. Myrdal has suggested that one of the sources of the U.S. (and anti-Marxist) theory of takeoff is Marx's argument in the preface to *Das Kapital*: A country that is more advanced industrially merely shows one not so advanced the image of its own future.[2]

No Blueprint for Development

According to this theory, associated with the name of W. W. Rostow, there is only one model of development—that of the industrialized countries—and growth passes necessarily through the famous five stages: traditional society, preparation for takeoff, takeoff, maturity, and the consumer society. Various stimuli can lead to takeoff: technological, as in Britain at the start of the Industrial Revolution; political, as in Japan at the time of the Meiji; or economic, if the price is right and the market opens up. These five stages are the same for all countries, no matter when they embark on industrialization.[3]

The simplicity of this theory made it very attractive to some economists, but many weighty arguments were also raised against it, especially by historians and anthropologists. A. Gerschenkron, for example, demonstrated that in Europe, industrialization in the laggards like Germany and Russia was fundamentally different from what happened in Britain. It is easy to scoff at Rostow's five stages that all countries are supposed to respect, yet listing the prerequisites for maturity is less important than understanding the nature of the obstacles to development. Gerschenkron's historical perspective and his analysis show conclusively that the paths to development are many and varied.[4] True, the laggard countries can benefit from the experience of the front-runners, but their futures cannot necessarily be predicted from what happened to the first to industrialize. Each case differs from the original model, so that each one—far from being an exact copy—creates its own version.

All the theories of development have nevertheless been influenced by the notion of takeoff, even if only so as to attack its deficiencies. First for economists, then for historians, the idea of takeoff probably helped in understanding more clearly some components of the Industrial Revolution by clarifying what happened at a given moment to alter the old rules of the economic game and so to cause, for a very brief period, takeoff and then rapid growth. This is the moment when the rates of growth of GNP and of investment part company.

To be fair, quite apart from the failure of the theory of takeoff to explain the peculiarities of different national experiences of growth, its methodology has been misused in applying it to modern developing countries as if the model, which might possibly hold for the preindustrial economies of Europe in the seventeenth and eighteenth centuries, had universal validity. If we adopt this approach, we overlook three facts the theory has always played down, if not totally neglected.

For one thing, today's developing countries, most of which have been under the close domination of the older industrial nations, are not in the same state in economic, social, and cultural terms as preindustrial Europe. The current changes are happening in exactly the opposite circumstances to those in which the first generation became industrialized. As S. N. Einsenstadt has shown, rapid urbanization, new needs and expectations, modern mechanisms of administration and politics, and the interdependence of international trade and debt all impose their constraints and effects before the economic and cultural base has been solidly established. Modernization occurs by fits and starts. It does not spread evenly through society if the ground is not prepared; it brings about a juxtaposition of very unequal sectors that have little contact with each other and is accompanied by "breakdowns of modernization" and greater polarization rather than the creation of more equitable income distribution.[5]

Second, the theory underestimates the importance of obstacles and resistance in the evolution of industrialized societies and glosses over the imbalances, distortions, and upheavals that growth generates in developing countries. By emphasizing the "positive" aspects of progress at the expense of "traditional" values, it offers a linear view of history in which the process of modernization is necessarily a change for the "better" (the stages of maturity and the consumer society, according to Rostow).

This transition, with all its attendant resistance, setbacks, and suffering, is nevertheless history as it is lived and made by human beings and not as it is conceived by economists. Has the mirage of the successful European or U.S. model made people forget the price paid (for example, in social conflicts as well as wars and economic crises) and the time it took to reach maturity? We are impressed by the entrepreneurs—the captains of industry, the great bankers, inventors, and engineers—who were responsible for the spectacular growth produced by industrialization, but we tend to forget the exploitation of women and children, the migration of labor, the traumas, and the problems of adjustment that were associated with the slow emergence from the preindustrial era.

Last, in the theory of takeoff, technical change is by definition a given imposed from outside—without reference to the economic, social, and cultural characteristics of each society—rather than a variable whose influence (capacity to spread and be assimilated) itself depends on these very characteristics. In particular, the theory overlooks a factor that has played a crucial role in the growth of industrialized countries: the widespread possession of basic technical skills as a result of mass primary education. The first Industrial Revolution was less the product of science and scientists than of technology and of craftspeople and supervisors with a skimpy knowledge of mathematics.

The theory of takeoff and its methodology would not have had significant repercussions had it remained merely a matter for academic debate. It did, however, inspire more or less directly the ideas and policies relating to development of both governments and international organizations, to the point that they lost sight of the most important thing: the human dimension—with all that means in terms of hesitation and resistance—in the interaction between technical change and social development and, linked to it, the incompressible dimension of time. The transition from the preindustrial stage to pseudo-maturity is a long process; to acquire the knowledge and the techniques essential to operate an industrial nation takes more than one generation.

Moreover, technical change is no longer what it was for the first generations of the Industrial Revolution. It now depends on a close, almost organic link between science and technology on one hand and industry on the other and therefore presupposes an even more sophisticated preparation. Indeed, the most decisive change in the history of industrialization is the industrialization of research in science and technology. In the nineteenth century, new institutions designed specifically to produce and market goods and services on a large scale grew up alongside the expansion of industrial capitalism. In the twentieth century, new institutions have been designed specifically to generate new technologies on a large scale: The laboratory rather than the large factory typifies modern industry. Furthermore, the Information Revolution is constantly raising the intellectual demands of the industrial system as the new technical paradigm creates yet another frontier, this one separating the *mastery of production* from the *mastery of use* of high-technology products and processes.

The consequence is that most developing countries are having to cope with a new set of obstacles in their attempts to modernize. In the nineteenth century, the preindustrial economies were able to embark on the process of industrialization using knowledge, know-how, and methods that were not heavily dependent on science. The steam engine was in use long before the theoretical principles involved had been understood, and the transition from blacksmith to mechanic able to repair a combustion engine occurred although only primary education had become widespread.

Today, by contrast, the industrial system depends upon an infrastructure of knowledge and technical skills that are closely linked to advanced scientific knowledge and that therefore require a high level of education and training. The management and maintenance of the system, and all the more the ability to update it, are in the hands of specialists. Access to scientific discovery and technical innovation is becoming increasingly difficult because they presuppose research teams, special academic and industrial institutions, and hence investments on a scale that for some programs may be out of reach of not only the largest firms but even some of the most highly industrialized countries. For most developing countries, hope of reaching Rostow's maturity stage has receded over the horizon, and the consumer society as it is understood in the most advanced industrialized countries can be no more than a mirage.

The Diversity of Circumstances

However strong the feelings of solidarity among the members of the Group of 77 (which number over 100), of the nonaligned nations, of the Arab League, or of OPEC, the developing countries are not a homogeneous entity. The deficiencies characteristic of underdevelopment are not of the same severity everywhere. There is not *one* Third World, there are *many*, and the degrees of underdevelopment are no less varied than the degrees of development. From direst poverty to real possibilities of catching up, the diversity and the differences count for far more than the similarities and affinities.

All efforts to classify or group countries will inevitably fail to account properly for the present situation or for the future. Only one thing is obvious: Third World countries measure themselves more against the industrialized countries than against each other, yet at the same time their distinctive differences—as they now are and as they evolve—must be taken into account in order to understand the role that science and technology can play in their development.

From the economic perspective, the potential capacity of each of these many Third Worlds to control its own development depends on the size and rate of increase of its population, the natural resources available, and the level of education and skills in the labor force that allows income to grow. From this standpoint, we can group the range of national and regional situations loosely as follows, all the while recognizing that there are many variants, especially with regard to political organization and to the strategies adopted toward industrialization in some cases and agricultural development in others:

- Countries where the capital resources, participation in world trade and industrialization have permitted takeoff;
- Countries where demographic pressures and the low level of mobilization of natural and human resources jeopardize development efforts;
- Between these two extremes, the oil-producing countries, some of them subject to strong demographic pressures, others having a population growth not out of line with their oil reserves.

During the last thirty years, there has been an undeniable improvement in the rates of industrialization in some Third World countries. Their exports to industrialized countries have grown by 15 percent in real terms, although they still account for only a minute part of the market, and the share of the developing countries in total industrial value added is growing extremely slowly. On closer inspection, it appears that four countries account for most of that growth: Brazil, Mexico, Argentina, and India. Only three countries export more than $1 billion manufactured goods every year: Brazil, Hong Kong, and South Korea. Tiny Hong Kong alone accounts for one-fifth of the exports of manufactured goods from the Third World, although this is thanks more to goods in transit than to actual domestic output.

Moreover, these exports are limited to a narrow range of products: clothing, textiles and shoes, metal and mechanical goods, wood products and furniture, and food. As for more sophisticated products (engines, electrical equipment, spare parts, microelectronics, and components), the major part in these developments has been played by the multinational firms and holding companies that have over 16,000 subsidiaries and branches in the same countries: Brazil, Mexico, India, South Korea, Hong Kong, Singapore, Taiwan, Pakistan, and the Philippines. Recently (1982–1986), a slight growth has been noticeable in exports of manufactured goods to the United States and, to a lesser extent, Europe. A substantial part of this growth is due to multinational firms and to products related to electronics, though the competitive position of a few countries would appear to be more broadly based.

Annual economic output in the newly industrialized countries (Brazil, Mexico, South Korea, Hong Kong, Singapore, and Taiwan) taken together grew faster (8.4 percent) than in the industrialized countries (4.8 percent) between 1964 and 1973; this rate fell to 5.9 percent between 1973 and 1982, while the rate in the industrialized countries fell to 2.1 percent. The NICs' share in the total gross domestic product of the whole noncommunist world virtually doubled between 1964 and 1982, rising from 3.6 to 6.3 percent, whereas that of the most industrialized OECD members fell from 72.3 percent in 1964 to 64.6 percent in 1982. As noted, most of the increase in exports of manufactured goods to the industrialized countries can be attributed to the NICs, mainly occuring in nontraditional products related to radio, television, and telecommunications equipment.

It is possible that in the future other countries will manage to join the ranks of those (like the NICs already) able to contribute to the design and marketing of new products related to the Information Revolution. This accolade is not, however, likely to be much more widely distributed between now and the end of the century because few nations fulfill the conditions that lie behind the NICs' technological success. The vast differences among the developing countries have been exacerbated by the effects of the oil crises: The process of industrialization has been speeded up in some of the oil-producing countries, whereas the rise in oil prices has substantially worsened the indebtedness of those without oil. None of the scenarios for the future can be very optimistic about the likelihood of the *majority* of developing countries being able to industrialize rapidly. As Jacques Lesourne has written: "All the roads leading out of underdevelopment are steep and some lead nowhere. So, just as in races where the slopes are too steep, the Third World team will go on falling behind during the next quarter-century."[6]

If $2,500 per capita annual income is (arbitrarily) reckoned to be the threshold of development, the inhabitants of the countries joining the takeoff group will make up 12 percent of the world population (the total in the group rising from 470 to 760 million). At the other extreme, if the poverty line is set at $300 per annum, the percentage of the world population below that level will fall from 32 to 28 percent—but demographic change will prevent any cry of victory because in absolute terms the numbers living in misery will increase, from 1.28 billion now to 1.65 billion.

Between the areas of desperate poverty (South Asia and tropical Africa) and the countries or regions that have some hope of improving their positions (Latin America, the Far East), there are many degrees of underdevelopment. We should not just talk about the "fourth world"; we should recognize that, just as there are underprivileged groups in rich countries, there are considerable disparities in the growth prospects of these countries taken individually. According to the most moderate scenario (provided by OECD's program *Interfutures*), the forecasts suggest categories that should not be seen as a strict classification but more as a way of highlighting typical situations (some countries may fit in several categories):

1. Countries where industrialization is quite advanced and the economy is becoming more diversified. These can be divided into two separate subgroups: (i) two medium-sized Asian countries, South Korea and Taiwan, plus the two city-states of Hong Kong and Singapore; (ii) the largest Latin American countries (Argentina, Brazil, and Mexico). Thanks to their small size, the countries in the first subgroup are saved from having the regional disparities found in the others, which are characterized by a population that is mostly very poor, vast backward regions where there are occasional periods of actual famine, yet at the same time an advanced industrial, scientific, and technological infrastructure producing some goods able to compete in world markets.

2. A more disparate group of countries where industrialization is increasingly important, such as Algeria, Venezuela, Malaysia, the Philippines, Pakistan, Nigeria, Kenya, and Côte d'Ivoire. Again these fall into two subgroups: at one extreme, those that could become industrialized

nations if circumstances were favorable; at the other, those with far less chance of escaping from underdevelopment by industrialization. Algeria, Venezuela, and Côte d'Ivoire probably belong in the first subgroup; Pakistan, Kenya, and Malaysia in the second. In most of these countries, however, agriculture will remain a crucial activity, often much more significant than industry in the development process.

3. Countries whose growth prospects depend primarily on natural resources. These, too, fall into subgroups. Included here are countries with an important market position in one or more ores or primary products (Saudi Arabia, the Gulf states, and other OPEC members for oil; Zaire, Zambia, Chile, and Peru for copper; Thailand, Malaysia, Bolivia, and Indonesia for tin; Jamaica and Guinea for aluminium; Ghana for cocoa), plus countries with too small a share of world markets in several products to be able to undertake rapid industrialization (Tanzania, Paraguay, Ethiopia).

4. Very poor countries with few natural resources and little prospect of industrialization, where improvements in agriculture are all the more needed than in the other groups.

5. The two continent-countries, India and China, where regional differences are blurred by their size, political unity, and common cultural heritage. Because of their size, they have features of all the other categories. Brazil could also be included here: Like India, it is simultaneously a developing country in its vast rural areas and also a major industrial nation, producing and exporting high-technology products, though Brazil does not suffer from the same problems of overpopulation. In China, as a result of its efforts to control population growth and stimulate agricultural production, per capita incomes in the rural areas are rising rapidly while the pockets of industrialization are only now beginning to expand.[7]

This classification is of course derived from scenarios of main trends. These trends are unlikely to change significantly before the end of the century, yet it is still possible that some countries within each group will be able to alter their prospects—for the better or worse—by the development strategies they adopt. In any case, it is quite clear that for most of them, agriculture rather than industry will remain the dominant activity. Consequently, any strategy that favors industry at the expense of agricultural output will normally threaten to accentuate the internal distortions and hence will jeopardize the chances of reestablishing a balance between the rate of population growth and the available natural resources.

The Noneconomic Factors

Last but not least, and whatever the scenario, it must be stressed that the obstacles to development do not involve simply the availability of natural resources stacked against demographic pressures. They depend as well—perhaps above all—upon social structure and the stability of the political system. Lack of natural resources and population size may by themselves determine the poverty line, but the nature of the social organization and of the political system defines the limits of a country's ability to mobilize its human and financial resources. Social and cultural imbalances, political instability, and inappropriate economic policies can slow or halt the process of development.

Even so-called developed countries are not protected from making the wrong economic choices or from political upheavals that lead along the road to underdevelopment. The case of Poland is illuminating. Quite apart from the political problems, the massive imports of Western technologies in the 1970s merely exacerbated the country's difficulties. At that time the USSR and the Eastern bloc were waking up to the technological gap between themselves and the West. The new scientific revolution set in motion by cybernetics was supposed to give a new impetus to all the communist economies. With the aim of modernizing their productive capacity, they were encouraged to invest heavily in the most recent technologies, raise spending on research, and increase the level of technology transfer from the West.

In Poland, the New Development Strategy soon encountered an insurmountable obstacle: To adopt the new methods and to increase the research effort meant first undertaking reforms unacceptable to the centralized planning and management systems. Investment levels were set far too high, while the rigid planning system, inappropriate organizational arrangements, and lack of skilled labor prevented the plans from being implemented. More and more bottlenecks developed, contributing to economic collapse and leaving machines, spare parts, and raw materials unused or unusable. The system thus seems to have been incompatible with massive doses of technology transfer because it prevented recourse to certain policies essential for technology transfer to succeed.[8]

Strictly economic definitions of underdevelopment therefore give a distorted view. They do not take account of the differences in development arising from such factors as the degree of political stability, how authoritarian the policies are, the class structure and the occupational structure, landownership patterns, skill levels in rural and urban areas, the investment in education, how widespread technical knowledge is, how long the universities have been established relative to other countries, the quality of training in the public and private sectors, and the deficiencies of the central administration—what Myrdal calls "the soft state," typical of all the developing countries, that allows corruption and arbitrary rule to flourish undisturbed. The list of not strictly economic factors is endless; in combination, in ways that are unfathomable yet critical, they determine a country's development chances.

This is why, after observing the development process over three decades, the best specialists in the field, like Gunnar Myrdal and Albert Hirschman, criticized the narrowness of the purely economic approach and argued that just as the explanation of underdevelopment cannot be reduced to quantifiable factors, the struggle against underdevelopment is not merely a matter of finding economic remedies. Myrdal always insisted that the problems of developing countries derive above all from non-economic factors such as patterns of behavior and the nature of institutions. He added that although few economists would now disagree with this view, not many would draw the logical conclusions:

> In presenting their concepts, models and theories, economists are regularly prepared to make the most generous reservations and qualifications—indeed, to emphasize that in the last instance development is a "human problem" and that planning means "changing men." Having thus made their bow to what they have become accustomed to call the "non-economic factors," they thereafter commonly proceed as if those factors did not exist.[9]

The shortsightedness of economists would not have mattered had it not affected the decisions taken by those with power in the developing countries.

3 A Basic Discontinuity

The history of the Industrial Revolution shows that the spread of technology internationally is a factor in catching up, and no one country can retain the monopoly over technical progress forever. Britain produced more steam engines in the midnineteenth century than all the other countries taken together, yet it soon lost its technological superiority. Although Britain led the world in the first stages of the Industrial Revolution (textiles, steam engines, iron and steel), the country fell further and further behind continental Europe and the United States in the subsequent stages (electric power, motor vehicles).

The latecomers in the race to industrialize certainly enjoy a comparative advantage in that there is a pool of available technologies they can copy or improve without having to take on the risks and the costs borne by the pioneers. There is no lack of examples of countries that, starting by copying others, achieved far more rapid rates of growth than their technological predecessors, such as the United States in the nineteenth century, Japan in the midtwentieth century, and more recently, the newly industrialized countries. Nevertheless, there must be something to build on: a well-established industrial base, an adequate pool of technical skills, an expanding domestic market—in short, a level of income and an absorption capacity high enough to permit takeoff or a fresh start. Moreover, we should not minimize the importance of an unquantifiable factor that even economists are obliged to take into account: the determination to develop. This explains why the rates of growth in Western Europe and Japan were far higher than in the United States after World War II, in spite of the enormous destruction and loss of life.

Nonetheless, although there is much greater global interdependence now than a century ago, with a larger volume of international trade being moved more rapidly, there are some areas where there can be no question of catching up. The very spread of new technologies establishes greater disparities than existed among European nations at the beginning of the

Industrial Revolution. The concepts and methods required to run or have an impact upon the world's business are becoming ever more sophisticated and consequently difficult to master, demanding increasingly specialized technical expertise and qualifications.

Until the middle of the twentieth century, the major technical innovations of the Industrial Revolution were based on existing knowledge rather than on work at the leading edge of science. As industrialization progressed, the European countries (and later the United States) were able to catch up with Britain all the more easily because it was not necessary to rely on science in order to introduce the advanced technologies of those days into the productive system. Intellectual capital—a labor force highly skilled in science and technology, plus institutions designed specifically to generate new technologies—did not then play a crucial role, and the resources needed to be part of the stream of innovations were well within the means of any industrial nation, large or small.

From Crécy to Hiroshima

The prerequisites for the mastery and use of the technical system changed once and for all, well before the Information Revolution, with the advent of nuclear power. The discovery of nuclear fission, leading to the making of the atom bomb, can be seen as the real turning point. Margaret Gowing has argued that this marked a watershed such as is rarely observed in the history of science, and indeed it soon marked a fundamental break in the history of civilization.[1] It is hard to find a term strong enough to convey the radical change in the power relations between and among states brought about by nuclear weapons, because nothing in human history could ever be the same as before. Not only did war and peace now mean something completely different, under the permanent threat of global destruction. Beyond the new power relations, a barrier appeared that most countries could not cross— one erected by capacity in science and technology.[2]

The same might be said of the introduction of gunpowder in the West, but it took far longer for cannons to become widespread—several centuries—than for the members of the "atomic club" to equip themselves with nuclear arsenals, which happened within a decade. Cannons are first known for certain in Flanders around 1314, but they did not become common until a century later, thanks to improvements in gunpowder. Before that, the new weapon created more noise than actual harm. At the battle of Crécy, the English guns merely "stupefied" the French troops of Philip VI, according to Jean Froissart's account.[3] Not until the fifteenth century was there development of heavy artillery, for use in the field or aboard ships, as well as of the harquebus, in effect a miniature cannon.

It is true there was henceforth a new distribution of military might—

only rich nations could afford the enormous costs of the new style of warfare. Artillery and guns led to the rise of states at the expense of independent cities and towns because of the vast expenditures required for defense. Braudel notes that in the sixteenth century the powder needed to defend Venice cost at least 1.8 million ducats, which was more than the annual income of the city authorities.[4] The new arms did not, however, remain the monopoly of the major powers of the time, but spread throughout the world, partly via European mercenaries. Skill in using the weapons long counted for far more than their number and size, which explains why there were victories on both sides in the wars between Europeans and Moslems or Turks. All in all, gunpowder did not fundamentally change the relationships among nations or the frontiers between the major cultural areas, although Europe probably benefited most, especially in the conquest of the New World.

By contrast, nuclear weapons—bombs, missiles, and their electronic guidance systems—are and will remain in the hands of a very small number of countries. Whatever the risks of nuclear proliferation, there will never be worldwide access to these weapons. True, it is just possible to imagine a situation in which terrorists, acting as individuals, could threaten to contaminate food or water supplies with radioactive substances; it is harder to imagine independent terrorists using a nuclear bomb to back up their threats—that is the radical change brought about by nuclear fission. Only states can muster the enormous scientific, technological, and industrial resources required to develop, manage, and safeguard these weapons. Above all, a nation that chooses to become a nuclear power must face the fact that this invites the possibility of total destruction.

Moreover, widespread use of nuclear energy for peaceful purposes is equally unlikely. Few countries will ever be in a position to design and build nuclear power stations; and merely to use them, without having taken part in their construction, means lasting dependency upon foreign technical assistance. Inequality in the capacity to undertake scientific research, and consequently to exploit the new technical system derived from it, is a given of the modern world that very few countries can hope to overcome.

The nuclear factor has helped accelerate the pace of technical change, which is becoming increasingly capital-intensive. The innovations now created, unlike nuclear weapons or energy, tend to spread rapidly throughout the world, yet the investments and institutional backup required mean that only a handful of countries—sometimes a handful of firms—are able to produce them. These innovations are responsible for the growth of new industries, which depend directly on intellectual capital for both the design and marketing of their products.

Competition among these industries is conducted in the context of the global economy, and battles for commercial survival often concern the definition and adoption of standards (for example, the often intangible technical systems such as SECAM versus VHS for color televisions, or the IBM personal computer). The most obvious result of these developments is not simply that very few countries are directly engaged in the competition, but also that the rest cannot acquire the know-how of these new industries by nationalizing them: It is possible to take over the physical plant of a firm but not the invisibles that are the source of its technical and commercial success.

The High-Tech Empire

The pace of technical change has been so rapid that new vocabulary is required to label the most advanced techniques and products in the ferment of new technologies—hence the creation of terms like "high tech" and now "ultra tech."[5] Close links with science, mastery of the technological factors that determine successful production and distribution, trade on a world scale—these are all areas in which the investments and risks are enormous, strategic importance is great for governments as well as firms, and products and methods involved are rapidly superseded.

The National Science Foundation puts a product into this category if it is made by a firm that employs at least 25 scientists and engineers for every 1,000 employees and spends at least 3.5 percent of its turnover on research and development (R&D). The Department of Commerce tries to be more precise and bases its definition on an input-output analysis of total R&D spending on a wide range of individual products. Thus an aircraft belongs in the high-tech category not only for its airframe but also for the manufacture of various components supplied by specialized subcontractors, including items like tires. According to this definition, high tech covers the ten most research-intensive sectors: missiles and satellite launchers; electronics and telecommunications; aircraft and aerospace; computers and office equipment; certain modern conventional weapons; medicines, vaccines, and hormones; inorganic chemical products; scientific instruments; advanced engines and turbines; plastics and synthetic fibers.

OECD for its part defines high-tech industries by examining the ratio of R&D expenditures to production. Its indicator is an average weighted by each industry's output as a proportion of total output in the eleven OECD member countries with the highest investment in R&D. High-tech industries are ten times more technology-intensive than the rest: aerospace 25.6; computers and office machines 13.4; electronics and components 8.4; pharmaceuticals 4.5. The comparable average is less than 1.7 for chemicals, automobiles, mechanical equipment, plastics, and oil refining, and 0.4 for other industries.[6]

These definitions fail to keep track of the progress of high tech. Most are in fact based on statistical categories established before the advent of the new technologies. Computers, for example, are included in the same category as typewriters, and quartz alarm clocks are classed as scientific instruments. Moreover, the level of aggregation is too high to be able to pick out the high-tech areas within traditional industries, even though these are increasing: The automobile industry is still in the "medium intensity" category even though vehicles are increasingly designed by computers and produced by robots. Furthermore, producers should be distinguished from consumers (this is essential for telecommunications and the media) and—above all—products from processes, since the refinement of software to meet the needs of industrial production and the application of computers to management have entirely altered the way industry and the economy function.

It may not be easy to define the domain of high tech, but it is nevertheless clear that it is playing an increasingly important role in the economies of the industrialized countries (it is the most dynamic sector, with record-breaking growth rates) and accounts for an ever larger share of international trade. In the United States, the electronics industry has become the major manufacturing employer; the turnover has now exceeded that of the automobile industry. The volume of trade in electronics has become one of the main indicators of a country's overall competitiveness.[7]

The United States is still the giant, dominating the rest of the world, but the U.S. share of the OECD market declined from 30 percent in 1960 to 24 percent in 1983, while that of Japan grew spectacularly. Several European countries (United Kingdom, Germany, France, and Switzerland) control almost one-third of this market, with some very strong sectors (30 percent of the world market in pharmaceuticals, which is more than the share of Europe in world gross output) and some in which they are ahead of the United States (nuclear power, satellite launchers).

International trade in high-tech products is generating increasing tensions in international gatherings among the industrialized countries, which are prepared to do almost anything to protect the growth of their fledgling industries. There is a fine line between restrictions imposed in the name of national security and outright protectionist measures to limit the access of foreign competitors to leading-edge products—a line easily crossed not only when the protagonists are in opposing political camps but also when they are allies. At meetings of the General Agreement on Tariffs and Trade (GATT) and of OECD, the U.S. representatives constantly maintain that the direct and indirect state aid given to European industry constitutes subsidies or protectionism. Meanwhile their adversaries (Europeans, Japanese, or Brazilians) respond by pointing to the help given to U.S. firms via Pentagon orders for new technologies that—thanks ultimately to military research—can later be launched on the civilian market.

The example of semiconductors is illustrative: France and Britain reacted to U.S. superiority in this field—a status encouraged by the vigorous support of the Department of Defense and the National Aeronautics and Space Administration (NASA)—by invoking the argument of a "vital or strategic technology" to justify state intervention in reorganizing and subsidizing industries aimed basically at supplying the public sector. While the U.S. firms worry about the decline of their position vis-à-vis Japan in electronics and Europe in aerospace, the Europeans are anxious to improve the quality of their electronics industries in order to confront the double challenge of the United States and Japan. Behind these "trade-related issues" lie strong protectionist tendencies, often disguised as concerns about defense or vital technologies, that some observers feel should be thwarted.[8]

The design and development of new technologies require enormous capital resources, a close link between science and technology (and therefore between universities and industry), and a large pool of highly qualified scientists, engineers, and technicians. Industrial research is becoming increasingly expensive: It is estimated that one research worker, plus support staff and equipment, costs on average $100,000 per year in the OECD countries.[9] The "smaller" a country in terms not of physical size and natural endowment but of financial resources and highly trained workers, the more limited the range of scientific fields and branches of industry in which it can hope to operate successfully. An aggressive research policy—in order to get well ahead of competitors—is increasingly costly, all the more so because the creation of a substantial part of the new technologies is linked to military research, and hence public authorities alone can take on the risks and the expenditures. Most countries, even if industrialized, are restricted to defensive policies only.

Military R&D can certainly appear to be a constraint on, if not actually detrimental to, the general well-being of the civilian economy; indeed, many U.S. observers find in this the explanation for the declining rate of productivity growth of their country compared with Japan or Germany. Nevertheless, it does act as a considerable stimulant to the training of researchers and for R&D activities in general, some of which have spinoffs that provide a comparative advantage in the medium or long term. In any case, the large size of the capital outlays involved explains why the new technologies have led to a tremendous concentration of R&D efforts at all levels (nations, institutions, technical area), and it gives some idea of the enormous gap separating the most industrialized nations from all the others, especially the developing countries.

- *Nations.* The OECD countries—which aside from the former Soviet Union account for the majority of the world's R&D—spent almost $192 billion on R&D in 1985. Half of this sum was spent in the United

States alone, one-third in the European Community countries, one-tenth in Japan, and less than one-tenth in the remaining countries. The bulk of the R&D is conducted in five OECD member countries (United States, Japan, Germany, France, Britain), which account for 86 percent of the total effort in the OECD area, measured in terms of expenditures and numbers of researchers. If Italy and Canada are added to the first five, the proportion rises to 91 percent.[10] Certain categories of R&D are so expensive (space research, development of "fifth-generation" computers) that they are beyond the reach even of most industrialized countries, hence the increasing tendency in Europe to organize cooperative research programs, such as ESPRIT (for computer sciences) within the EC and Eureka (industrial pre-competitive research) covering all of Western Europe.

• *Institutions.* The U.S. Department of Defense alone controls 10 percent of all the expenditure on R&D in the OECD area, or the equivalent of the joint expenditures of Italy, Canada, the Netherlands, Sweden, Switzerland, Australia, and Belgium (or of the total research budget of Germany). Ten public and private institutions (government agencies and multinational companies) spend almost one-third of the total R&D budget of the OECD area. The aggregate research spending of firms like General Motors, Ford, IBM, AT&T, General Electric, United Technology, Boeing, Kodak, ITT, and Dupont ($9 billion) is $1 billion more than the research budget of countries such as France or Britain. Universities like MIT (Massachusetts Institute of Technology), Caltech (California Institute of Technology), or Stanford have research budgets equivalent to one-third of the national research budgets of, for example, Austria or Norway.

• *Technical area.* In the United States and Britain, half of public R&D spending goes to defense; in France and Sweden the proportions are almost as large. Among the specific goals and policies aimed at research and information in the OECD area, defense and space programs account for 44 percent; telecommunications, transport, energy, and urban renewal 21 percent; agriculture and manufacturing 15 percent; health and social services 11 percent; and basic research 10 percent. The higher-education sector in the United States alone spent $11.3 million on R&D in 1983, which was almost as much as the total R&D budget of France ($13.13 million) or Britain ($12.55 million). Private-sector R&D efforts tend to be concentrated in manufacturing, particularly engineering (space, electrical and electronic, computers and machinery, vehicles, and marine) and chemicals (petroleum and coal byproducts, plastics, and medicines).

The increasingly fierce competition in this area, the high cost of research, and the need to reach as large a market as possible as rapidly as possible together explain the emergence of a new phenomenon: cooperation agreements, joint ventures between competing firms—sometimes in the same country, sometimes in different countries—to pool their scientific resources and make a concerted attack on either all or part of an R&D program. Even the biggest firms in the United States and Europe are discovering that they too are subject to the limits that caused European governments to undertake joint research projects. In the same way that the new technical system diminishes the sovereignty of the nation-state, so it undermines the autonomy of firms, which become multinationals not simply to sell their products through their subsidiaries in many countries but also to set up networks for arranging and implementing joint research programs.[11]

High-technology industries account for only a small proportion of total manufacturing output (roughly 11 percent on average) and a slightly higher proportion of trade (16 percent), yet they have recorded the fastest growth in the whole OECD area since 1970, measured in terms of output, domestic demand, imports, and exports. These new technologies give us a glimpse of what the industrial landscape will be like in the twenty-first century: The entire industrial system is being shaped and manipulated by the "big spenders" on R&D, which compete through innovation and hence alter the basis of international trade as well as reduce the room for maneuver of the less big, let alone the small.

Such were the findings of a meeting organized by OECD in Helsinki in 1986 on science and technology policy in small industrialized countries.[12] The smaller European countries may rank among the most advanced nations, yet in fact they struggle to stay in the race for innovations. The poorest among them (Ireland, Greece, Portugal, Yugoslavia) have even less in the way of spare resources to spend on R&D and hence find it hard just to stay in the race. In this regard, the "less developed" members of OECD are no different from Third World countries. A report by the European Parliament on these countries' expenditures on new technologies—information technologies, biotechnology, new materials—showed that their spending was so small as to be essentially unproductive, quite apart from their poor research structures and obstacles to entrepreneurship and innovation.[13]

All things are relative, of course: The "less developed" OECD members do not belong to the same category as the "developing countries" of the Third World. They do not suffer from famine or epidemics, and their population growth does not outstrip agricultural production. Nevertheless, faced with the exceptional concentration of R&D efforts, even some highly industrialized countries wonder whether they are going to be pushed off the field where the battles of international trade are to be waged in the future. This is all the more true of the less advanced OECD nations, which "are concerned that, having missed one industrial revolution, they are not left behind by the next."[14] It is therefore understandable that the Third World countries—defined above all in terms of their shortage of capital and abundance of labor, plus too little food—worry about their capacity to adapt to the impact of the new technologies, to control their importation, and to identify the ones they really need.

As far as R&D is concerned, most of these countries are simply not in the running. To echo Marx's famous exclamation about the weakness of the Olympian gods compared with the power of the institutions that in his eyes embodied the vastness and dynamism of nineteenth-century industrial capitalism:[15] How can the Third World countries' efforts in science and technology be anything but insignificant compared with those of the multinational firms, whose R&D budgets are larger than those of many industrialized countries?

The Experience of Latin America

In the 1960s, the several Third Worlds began to generate a vast literature on the importance of science and technology and the actual or potential links these resources might have with economic and social development. The way the thinking on these matters evolved in Latin America is especially revealing, because that was where the liveliest debate occurred—probably because some countries were fairly advanced, particularly as regards industrialization, but also because of the serious crisis provoked by their process of industrialization. The debate, with the hopes and disappointments it generated, helps to clarify the constraints hindering the application of science and technology to the problems of development.

The "social image" of science and technology—the way society has perceived the role and contributions of science, in particular its most active areas, to the process of economic and social development—has passed through four stages in the course of the last sixty years. After an "aristocratic" phase before World War II, there was a "scientistic" phase in the latter part of the 1950s and early 1960s, followed by a period when science and technology policies were drawn up (second half of the 1960s through the 1970s), and eventually, a stage of disenchantment and uncertainty that began to creep in toward the end of the 1970s.

Science the Aristocrat

Science was seen as the preserve of a small elite group of savants, entirely detached from society and social conflicts; these people were thought to be individualists, rather solitary, the luxury products of European culture, devoting themselves to research that was in no way related to local problems. Their activities were conducted in a handful of laboratories, and the scientists seemed to be an intellectual aristocracy isolated from the rest of society, their work ignored—sometimes even despised—by the great landowners and the first industrial entrepreneurs. Foreign geologists and mining engineers were responsible for most of the surveys of mineral resources. Only biomedical research, in the tradition of Pasteur (represented in particular by the work of Oswaldo Cruz in Brazil), played a significant role in changing individual and collective behavior by establishing the basis of policies for public hygiene and health via the first efforts at preventive medicine and vaccination. The earliest laboratories were in any case created by affluent individual scientists rather than set up with direct support from the state.[16]

In the 1930s, Latin America started a process of industrialization geared to import substitution, and this trend accelerated in the 1940s. Consumer goods previously imported started to be manufactured locally, and capital equipment began to be imported instead. The inconcinnity between the structural changes that occurred in countries like Argentina and Brazil and the survival of an entirely humanist and elitist view of scientific research was all the more marked in that the manufacturing sector grew extremely rapidly. In Argentina, for example, the value of industrial output overtook that of agricultural production in the second half of the 1940s, but this industrial growth was based largely on foreign techniques and technicians.

The more industry came to rely on science and technology, the more it had to resort to skills brought in from abroad. The universities were not yet sufficiently mature, nor indeed were they turning out a sufficient number of students, to train the scientists that industry required. The view of science as the domain of the "happy few" continued to prevail in the scientific community and remained the dominant image among most sections of society. This aristocratic notion of science incorporated the idea that scientists acted as a sort of relay for "superior Western culture": They were people who, through their links with specialists in advanced countries, would guide the modernization of the country's intellectual life rather than its productive activities.

The intellectual elite in Latin America was, however, little more than a "dominant/dominated" group, to use an expression of Alonso Aguilar, the Mexican economist. The scientists had close ties with the scientific community in the developed countries, while at the same time their concerns distanced them from the economic and social preoccupations in their own countries. Locally, they belonged to the upper class, yet lack of adequate institutions and resources made them heavily and often passively dependent upon European and North American science and scientists.

The Period of Scientism

During this period, covering the late 1950s and early 1960s, modernization and industrialization proceeded, but the impetus given by the import-substitution trend disappeared. The universities expanded rapidly, student numbers rose as the middle classes grew and prospered, and there was a wave of optimism about development in Latin America. This was the period of *desarrollo* (development) that encouraged dreams of rapid modernization of the economy as well as hopes of political emancipation. The image of the solitary savant, detached from the crowd, was replaced by a more modern view, of the scientist-cum-entrepreneur, working in a large laboratory and hobnobbing with politicians. The United States, not Europe, now supplied the model to be copied.

Nevertheless, Latin American science retained its elitist bias and continued to be considered an exogenous factor having nothing to do with the development process. Scientists were expected above all to be attached to universities, keeping their distance from the rest of society. True, scientific activities might impinge upon social trends via technical innovation, but it was felt that university research should not be too closely linked with business interests and even less with political ones. If the state had any role in the advancement of science, it was to provide support, mainly financial, for basic research, which was seen as a spontaneous generator of social change.

Two factors in these developments should be noted. At that time, everything seemed possible with the help of science; the optimism prevailing in the industrialized countries rubbed off on the developing countries: "The attitude to science and technology was shaped by confidence and hope. If the world had problems, it was because there was too little science, or the wrong kind of science, or because people did not know how to use it properly."[17] With more science, the right kind of science, with a more intelligent use of scientific resources, the process of development would happen more quickly. This optimistic view probably strengthened the social legitimacy of the scientific elites and their independence as well, but it did not lead to closer cooperation between universities and industry.

The only area in which there were the beginnings of cooperation was based on self-deception, as certain countries (such as Argentina and Brazil) dreamed of major nuclear programs with power stations—and who knows, eventually maybe bombs—similar to those of the atomic club. Most scientists had few illusions about the viability of these projects at the time. There were not enough specialists, the industrial infrastructure was inadequate, and the political constraints were too powerful (military regimes and pressure from the United States). The nuclear programs fell further and further behind and encountered numerous setbacks, but the scientists took advantage of the government projects to train their students and improve their labs, knowing full well that although they could not satisfy the ambitions of the military, it was at least an opportunity to raise the quality of the universities. Brazil might boast of having mastered the techniques to make enriched uranium, but economic factors prevented these from being exploited. It was not in Latin America but in India and China that these ambitions were fulfilled: Well-protected from political upheavals, their nuclear programs were given absolute priority and benefited from a generation of physicists and chemists trained in Europe and the United States before World War II.

The second factor was the liberal tradition of academic autonomy, which rapidly emerged as a means of withstanding abuses by, as well as pressures from, the political sphere. Lobbies of major landowners in collusion with army officers, frequent coups d'état, failed popular uprisings, and revolutions did little to encourage scientific institutions to flourish in Latin America. Whether or not they were committed supporters of the opposition, scientists suffered from the same repression as other intellectuals, and some were obliged to emigrate. As a hotbed of protest and a bastion against political pressures, viewed with suspicion and criticized by politicians, the university became a place of shelter where natural—though not social—scientists could pursue their research to their own ends.

The Growth of Science Policies

In the second half of the 1960s, more and more people in Latin America began to pay greater attention to science and technology for their potential as tools of social advance and national independence. Political parties, trade unions, and governments became interested in these activities, thinking they provided all the answers about modernization, and the debate about *dependencia* seized on the obstacles to technology transfer as fresh arguments in favor of interventionist policies. As this awareness grew, the example of the industrialized countries—which ever since the first Sputnik had been equipping themselves with institutions and programs specifically designed to identify and implement such strategies—was again extremely influential.

Other factors also played a part, as has been stressed by Joseph Hodara, former director of the Science and Technology Unit of ECLA (the United Nations Economic Commission for Latin America): the growing role of the state in the development process; the criticism in the industrialized countries of the negative effects of technical change, which in Latin America focused on the importation of technologies unsuited to local needs; and finally, in the international context, the politicization of international relations and the growing interdependence of the world economic system, the increasing numbers of new nations created as former colonies became independent, the mounting tone in East-West confrontations, and the influence of international agencies and intergovernmental organizations responsible for promoting scientific and technical development.[18] In a revealing move, the Organization of American States created a scientific secretariat in Washington along the lines of OECD's directorate.

These appeals to the state and interventions by international organizations and programs so as to strengthen scientific and technical resources were linked to growing anxieties about "technological dependence," which was seen as a factor leading to underdevelopment. Institutions were therefore created, laws enacted, and plans and programs drawn up with a view to implementing science and technology policies. The few relative successes (e.g., Mexico, Venezuela) do not alter the fact that the overall results were very limited. Those responsible for the policies tended to be indecisive rather than single-minded; they had inadequate financial and institutional support; and they had to cope with both extremely unstable political situations and the hostility of the multinational corporations.

Brazil was an exception, thanks to its vast natural resources and its long industrial experience. The public sector played a decisive part in encouraging and guiding investments. Interventionism led Brazil to undertake an ambitious plan for scientific and technological development, which would rely on increasing economic growth, training competent engineers at the military academies, and an energetic higher-education policy. The government took steps to control imports and provided direct support for local projects in industrial engineering and to build plant and machinery. In the energy sector, for instance, the national oil company secured a guarantee that from 1965 onward, 20 percent of the industry's materials should be supplied by Brazilian firms, to which the state offered loans on favorable terms. From then on, all the major capital programs were carried out under tightly controlled cooperation arrangements between foreign and local firms. Brazil thus used its industrial base to construct relatively independent national automobile and aircraft industries, followed shortly by computers.

The failures elsewhere, despite national differences, arose largely from similar problems related to political and economic circumstances. First, the state's power to act as a planning authority turned out to have been overestimated: The administrative machinery, political structures, and commitments—far from encouraging takeoff—simply reinforced underdevelopment. The policies and directives had no results; the state, itself a product of underdevelopment, was planning in thin air.

Second, the role and the influence of the multinationals, hitherto the sole mechanism of technology transfer and a key element in the economy, had been underestimated. The foreign firms had a practical monopoly of technical innovations and were rarely interested in sharing their know-how with the countries where they operated. Few countries were strong enough to challenge them. Unless they had an interventionist policy, aiming both to give foreign firms access to local raw materials and to try to achieve international quality standards locally, the host countries often found that importing technologies acted more as a hindrance than a stimulant to scientific and technical development.

Last, another factor had been underestimated: the unwillingness of the academic community to play along with the wishes of governments and politicians of dubious legitimacy and concern for democratic methods. Scientists attached all the greater importance to their links with the international community because they could not support the policies of the dictatorships. Thus, the notion of the independence of science and the demand to be able to do research for its own sake, like art for art's sake, unconnected with local concerns and the most urgent needs, became—if not a form of political protest—at least a way to escape from the pressures applied by military rulers.

Disillusionment

The pessimism about science and technology policies that can be observed nowadays derives in part from disappointment with the outcome of modernization and industrialization efforts. It was not easy to assess the impact of these strategies from the contradictory messages provided by apparent success stories like that of Brazil (exporting cars, arms, aircraft, and computers yet with 80 percent of the population left behind by progress, enormous regional disparities, and a foreign debt of $100 billion). The record indeed seemed bleak: the overall decline of manufacturing in Argentina; the isolation of Pinochet's Chile; the increasing poverty of the mass of the population in Mexico and Venezuela; the ideological confrontations and interminable civil wars in Central America.

Problems with Development—or with Rationality?

The same kinds of disappointments as in Latin America are to be found in other developing areas. The difficulties with development created suspicion about the "paradigm of rationality" that had inspired the economic and technological choices made by Third World countries.[19] In addition to the unbridgeable gulfs between nations, people had started to become aware of the costs of the ecological imbalances resulting from pursuing economic growth entirely through industrialization. This realization owed much to the MIT report on "the limits to growth" prepared for the Club of Rome, a document that helped to highlight the ambiguous contribution of some scientific and technological approaches to damaging the physical environments as well as the management of resources and the institutional structures of developing countries. The idea (almost an article of faith) that the modern sector of the economy should grow inexorably, transforming and taking over the traditional (if not actually antiquated) sectors in developing countries, was all the more susceptible to attack in that the industrialized countries themselves were beginning to recognize the ecological hazards in the model of industrialization they provided.

It is therefore hardly surprising if in most countries people started to question the good sense of the technological options hitherto chosen. Amilcar Herrera, for example, wrote: "The science and technology systems of the underdeveloped countries have proved themselves unable to produce national technologies in significant amounts."[20] In his view, the rush to modernize and industrialize had been the cause of major social upheavals, and the best way to make use of scientific and technical resources for development would be by adopting a completely new strategy based on labor-intensive methods for rural areas, trying to satisfy the basic needs of the population (food, housing, health, education), and preserving the heritage of traditional culture.

Similar views were held by Celso Furtado, the Brazilian economist who was planning minister until the military takeover and who ultimately returned from exile in France to become minister of culture under President José Sarney. Economic development is a myth, if it is taken to mean that the poor countries should simply reproduce the model of consumerism of the rich countries. In fact, this would introduce a deep rift in the productive system because there would be two very different technological levels. The more advanced one would supply the consumer demands of the most affluent sections of society almost exclusively, while the majority of the population would remain below subsistence level.[21]

In these circumstances, the developing countries became increasingly dependent upon the big multinational companies, which are able to impose and change the "basket" of consumer goods supplied. "Technical progress is no longer a problem of importing such or such a piece of machinery, but is rather becoming the problem of whether or not there is access to the flood of innovations being launched in the central economies."[22] By substituting their subsidiaries for local firms, the multinationals ensure that dependence is embedded in the local productive system, jeopardizing any attempts at autonomous development. It is therefore essential to control both the influence of the multinationals and the spread of new technologies to the peripheral economies of the Third World: "Increasing the consumption levels of the majority means encouraging the spread of existing products which are probably in a phase of rising productivity."[23]

Nevertheless, the debate is more inconclusive than ever between those countries that are relying on acquiring the most advanced technologies as the solution to the problems of underdevelopment and those that are banking on relatively traditional and "appropriate" technologies in order to create jobs and cope with the most urgent challenges. Should technology choices ignore the realities of the social and economic situation, or should the latter determine the former? The argument continues and sometimes takes surprising turns, as when Arghiri Emmanuel based his criticism of endogenous development models on a Marxist analysis. He maintained that no technology is tailor-made for the developing countries. The important factor for their social welfare and independence is the quantity of goods produced, not the number of jobs created in order to produce the good. Only the most advanced technologies create openings for technical specialists; hence, as the multinationals encourage efforts to train a specialized labor force by their simple presence as well as their recruitment, the "technological shortcut" for the Third World is provided by these firms rather than by modifying social structures.[24]

This unexpected combination of Marxism and capitalism indicates the uncertainties, if not confusion, of the debate about the contributions of science and technology. People argue as if the route taken by scientific and technological progress in the rich countries will necessarily lead to development in the poor countries, regardless of their initial situations. However, can the rest of the Third World really hope to emulate the handful of countries with a research capacity (and, even more so, the newly industrializing countries), prepared by their past—their cultural heritage, their industrial base, their social organization, and their interventionist policies—to take advantage of scientific and technological progress? Some want to catch up to the industrialized countries, others hope to compete with them, but the majority must first learn to stay afloat.

4 The Contemporary Technical System

Everything has to do with technique, Fernand Braudel argued, in that the human race, *homo faber* by nature, does not exist without material civilization.[1] However, while technology is an essential ingredient of development, it is not the sole ingredient. The technical infrastructure and capacity are merely indicators, along with others, and they obviously do not provide the definitive statement on collective welfare or social harmony. Yet they do at least help in assessing the relative level achieved by a given group of countries and the room for maneuver they have in trying to benefit from—or withstand the impact of—the innovations characteristic of the modern technical system.

In these terms, the industrialized countries are in fact the most "advanced," given their ability to master both the production and use of new technologies. Within the group there are, of course, leaders and followers—countries directly involved in generating new technologies and those more or less able to apply new techniques in pursuing their economic and social activities. And there are developing countries that, in spite of the disparities in other respects, stand out from all the rest in their ability to compete with the most advanced countries in exploiting certain components of the system.

The Idea of a Technical System

According to the historian of technology, Bertrand Gille, the complex interplay of relationships among the various forms of technical activity suggests that we should think in terms of "technical systems" evolving and replacing one another in the course of time. Each of the elements in the system needs one or more of the products of the other elements in order to function; the way these various elements complement and reinforce each other establishes the character of the new system at a given period and distinguishes it from its predecessors.

For example, at the start of the Industrial Revolution, the development of the steam engine was linked not only to advances in metallurgy but also to other technologies, to the point that they became interdependent. The steel industry used steam engines because of the need to produce ever stronger metal that would withstand high pressure and high temperatures, just as the railways depended on metallurgy becoming more sophisticated (higher-quality iron, improved techniques for lamination, brakes, and wheel-making). Yet this new technical system also needed the related progress in machine tools, telegraphy, and (equally important) new forms of work organization in order to achieve equilibrium. "As a very general rule, all techniques are in varying degrees dependent on one another, and there must be some consistency among them: the whole set of complementarities at different levels and among different branches of activity makes up what we may call a technical system."[2]

Ever since the beginning of the Industrial Revolution, technologies have been evolving and improving upon themselves so swiftly, with such enormous concomitant economic, social, and cultural changes, that it is impossible to separate the two strands: Each is both the cause and the effect of change. At the same time, one of the major issues of development everywhere—in the industrialized countries but even more so in those that are not—has become that of the compatibility between the nature of the new technical system and the receptiveness of the social structures in which the system's products and consequences will make their impact.

Societies are no more alike in their capacity to adapt to these changes than the technical system is static. The faster technologies evolve, the more they become the source of uncertainty about the future. Technical change not only constantly challenges the league table of development, but creates inconsistencies between the technical system on the one hand and the economic and social structures that it affects on the other. Increasingly, the interplay among social organization, the education system, and the levels of skill and industrialization, as well as the pace of change and the problems of adjustment generated by it, upsets the overall balance between the leaders and followers in innovation and adaptation.

Mechanization lay at the heart of the innovations that characterize the process of industrialization, leading to higher productivity, lower costs of producing goods and services, increased market size, and greater volume of trade. All these aims of mechanization depended on both technical and social changes. The result was the dynamism of industrial capitalism, which Karl Marx and Joseph Schumpeter found so "revolutionary." Technical change occurred by fits and starts, bringing about qualitative changes that in turn radically altered the previous equilibrium situation. This was indeed a process of "creative destruction," which, as Schumpeter said, did not mean increasing the total number of stagecoaches but rather replacing them with the railways. Capital growth is merely an element in the process; it does not provide the driving force for it.[3]

Mechanization obviously existed well before the Industrial Revolution; it developed during the Middle Ages and made remarkable progress during the Renaissance. Machines intended for industrial use, for metalworking and for manufacturing large objects, were built and used long before modern heavy industry came into being. However, mechanization faced two constraints until the eighteenth century that explain the slow pace of change, its limited spread, and the small range of social and economic repercussions it generated in the craft workshops, mills, and early factories where it was already to be found: the materials used for the machines (mainly wood) and the type of energy available (animals, human beings, or water power).

The substitution of metal for wood and, with the spread of the steam engine, coal for water brought about the growth of a new technical system. The transformations happened at increasing speed: After coal there came in succession electricity, the automobile and oil, aircraft, and nuclear power. The output of simple machines (symbolized by the combination of rod and crank, with the basic purpose of converting the application of a force) was considerably expanded by the creation of complex machines (symbolized by the motor, with the basic purpose of converting energy into work). The technical system created by the Industrial Revolution replaced the physical effort of human beings in production tasks on an ever greater scale.

This technical system, based essentially on harnessing energy and materials, was dominant in the industrialized countries until the middle of the twentieth century. It is now being replaced before our very eyes by a new system composed of all kinds of elements, ranging from computers to missiles, from robots to telecommunications, from biotechnologies to new materials. Systems of machines are gradually taking the place of complex machines for producing goods and services: The microprocessor (the "chip") is the symbol of this radical shift in which the purpose of the new machines is not merely to increase physical force but rather to extend and even sometimes replace the thinking functions of the brain.

From One Radical Shift to Another

The term "revolution," whether used in connection with technical or social change, has become so hackneyed that one hesitates to use it. As regards the information technologies, how far are we justified in talking about a complete break or a significant turning point? Are we at a sufficient distance to apply these terms to the technical transformations

we are witnessing? The question is all the more worth raising because there is no single frame of reference for analyzing the main trends within technical change. Historians, economists, anthropologists, and sociologists have all dealt with the topic in their own way, for their own ends, depending on whether they are more interested in the sources of invention, the determinants of innovation, the organization of production, the procedures, or the final products.

The choice of frame of reference, as well as of the period examined, affects the interpretation given to technical change. An anthropologist, for example—in this case Claude Lévi-Strauss—will see only two major revolutions in the course of human history: the Neolithic revolution, with the shift from nomadic hunter-gatherers to settled communities of farmers, and the industrial revolution.[4] Looked at from this angle, there were merely successive phases, involving greater or lesser social changes within the process of industrialization from its beginnings in Britain in the mideighteenth century, and it is still the same experience—whatever the scale of the transformations and the particular features of the various phases or means of industrialization peculiar to each country—that has occurred throughout the world ever since.

Within the Industrial Revolution, a historian of technology is likely to prefer to talk about evolution rather than revolution, because it is very rare for an invention to be the product of one person, one date, or one place. Invention seems instead to be a long-term process, which is both the result of the accumulated experience of several generations and the source of developments that have nothing whatsoever to do with the inventors. For instance, the history of the steam engine in fact began a century before Matthew Boulton and James Watt, and steam was not used widely by industry as a source of power until a quarter century after their inventions and the decisive refinements to the initial designs.[5] By contrast, an economist would not hesitate to talk about revolution with regard to technical innovations.

This difference is actually superficial, the result of what Schumpeter called a "difference of purpose and method only." As he so rightly said, "What we designate as a major source of invention hardly ever springs out of the current of events as Athene did from the head of Zeus, and practically every exception we might think of vanishes on closer investigation." From a historical perspective, an invention cannot be understood without reference to its remote beginnings—"it sums up rather than initiates."[6] Meanwhile, from an economic perspective, an invention backed by the market (which makes it an innovation) can at the same time be a source of major discontinuity. In passing from a microscopic concern with origins to a macroscopic concern with outcomes, the economist sees evolution as progressing by jerks and jumps, whereas the historian of technology sees in it primarily the continuity of developments.

We borrow from both these frames of reference in trying to bring out the characteristic features of the modern technical system, and we base our analysis on three concepts taken from physics: matter, energy, and information. The first two are old and need no comment; the third has joined the conceptual armory of the exact sciences more recently. We are talking here not about information in the sociologists' sense but of a *physical magnitude* that measures the amount of information contained in a message or transmitted by a signal. The concept of matter developed gradually from the beginnings of scientific knowledge; the concept of energy is more directly linked to the work of nineteenth-century physicists. The concept of information, as a measurable physical magnitude, dates from our own century with the work of C. E. Shannon, W. Weaver, L. Brillouin, Norbert Wiener, John von Neumann, and others; on it are based the theories of information, telecommunications, and encoding.[7]

By applying these concepts, we can class the operations involved in modern technology in three broad categories:

- Those that affect matter, whether living or inert, and move or reshape it or else cause it to undergo physical or chemical transformations;
- Those that pertain to energy and transform, transport, store, or dissipate it;
- Those that deal with information and, similarly, transfer, process, store, or destroy it.

These three categories are inextricably interlinked in most technical activities, although in any given instance there is normally one that predominates, while the other two are of secondary and less obvious importance. The main innovations that spread through the productive system can usually be placed unambiguously in one or another of these categories. For example, mastery of the transformation of heat into work dominated the first phase of the Industrial Revolution. The technical system then developed as a result of the mastery of electricity, based on the two extremes reflecting the standard division into "strong" and "weak" current—the former used to transfer energy, the latter to transfer information. In fact, both extremes were present (although the dominant one obscured the other) in Watt's energy-converting steam engine, perfected in 1788: His governor is really an ancestor of information technologies because it enabled the steam engine to use information about its own output to keep its speed regular (regulating the steam entering the cylinder).

In the history of technical change, three unbroken progression routes thus run together—moving forward at varying speeds at different times—in the development of the mastery of energy, of matter, and of information. At certain moments, progress in one of these is so much faster than in the others that it seems to put its exclusive imprint on technical change, but a close examination will reveal that this predominance does not mean the others are stationary. The nineteenth century appears to have been the century of energy. It was also the era of Joseph Jacquard's loom (another ancestor of information technologies, because of the use of punched cards), of the telegraph and telephone, and of the first radio links. The last quarter of the twentieth century may appear to be the era of information, yet it has seen advances just as great in nuclear energy, space vehicles and equipment, chemical synthesis, and biotechnologies.

Moreover, the progression routes of these three components of the technical system are not independent of one another; there is an interplay among them—advances in one contribute to the development of the others. The growth of electronics and of information technologies has increasingly facilitated the development of the others, not only with regard to energy and matter but also to the processes of living and cognition—calculation, modelbuilding, knowledge, and understanding. Two special features distinguish the contemporary technical system from its predecessors: the symbiotic relationship between science and technology, and the ever-increasing level of complexity. The most obvious likely effect of the latter will be steadily to displace the input of the human brain from the productive cycle. In this way, too, the contemporary technical system will add yet another dimension to the challenges of development.

The Symbiotic Relationship Between Science and Technology

The universality of Western science derives from its postulate that the laws governing the universe hold throughout all time and space. Its strength as an explanatory system is based on the success of this postulate; as Einstein expressed it in his well-known statement, "What is incomprehensible is that the world is comprehensible." Furthermore, the approach used by Western science gives it an additional operational power that is inseparable from its capacity to explain. The approach relies on a sequence of stages—formulating a theory that interprets a range of phenomena identified by observation or experimentation, making predictions based on the theory, checking the predictions by conducting experiments—and these are what make it possible for the scientific paradigm to advance. As Karl Popper has stressed, the possibility of refuting a theory by observation or experimentation is the very essence of scientific advance, because this leads to efforts to perfect or reformulate the theory and hence is both the source of and the prerequisite for progress.[8]

The only difference between the scientific experiment designed to test a theory and the technical artifact that puts the theory into practice lies in the ultimate purpose: Both are based on predictions derived from the theory; both are therefore capable of overturning the theory. But whereas refutation is the main aim of the scientific experiment, for the technical artifact refutation reduces the operational applicability of knowledge. In both cases, the theory must be reexamined. The stakes are not, however, the same, as is clearly implied by the terms "basic research" and "applied research": The first is about *knowing,* whereas the second is about *doing.*

The explanatory power of science is now so closely linked to its operational power that it is no longer possible to draw a sharp line between the two types of research. Even the conceptualization of the experimental method relies constantly and systematically on technology. Whether to build equipment for experimental purposes or to construct theories—for which the computer has become an indispensable tool—progress in basic research requires the most sophisticated products of technology. In chemistry and biology, for example, researchers are increasingly dependent on computers, not only to analyze chemical and genetic sequences more rapidly and more thoroughly but also to conceptualize new molecular structures and syntheses.

This symbiotic relationship between theoretical and operational, basic and applied science, between knowledge and know-how, reinforces its own success. For the last hundred years, the process of cross-fertilization has caused the pace of technical change to accelerate in the industrialized countries. Nevertheless, it would be an exaggeration to attribute rapid technical change entirely to this phenomenon, not least because of the importance of the social and economic context, in particular the growth of demand. The speed and direction of technical change do not depend solely on the logical framework set by technology—that of the scientific researchers, engineers, and inventors ("technology push"); they also depend on the market, determined by entrepreneurs and investors, to provide favorable conditions ("market pull").[9]

In addition, this symbiotic relationship between science and technology has developed so recently that there is still no clear answer to the political question that governments and organizations responsible for supporting scientific research must face all the time: What is the real impact of basic research as opposed to technology on economic and social development? Basic research, until the end of the nineteenth century, did not require substantial financial investment. State support remained all the more marginal because science was not in a position to guarantee any practical outcome, so that it was a matter of patronage rather than an essential budget item.

From the earliest days of *homo faber* to industrialization, the main source of technical progress was empiricism; the development of technical know-how only occasionally impinged on that of scientific knowledge, and rarely sufficiently to affect their independence from one another. Christian Huygens's application to clockmaking of Galileo's discovery that pendulums are isochronic stands out as an exception, an infrequent case of a technical improvement being directly inspired by science. Rather than guiding technical change, science was indebted to it insofar as it provided new topics for research and, above all, provided new means (via scientific instruments) for successfully conducting that research.[10] Practice came before theory: Just as compasses were made and used centuries before there was any scientific study of magnetism, so the mastery of the transformation of heat into work began and evolved in an entirely empirical fashion for a long time. Steam engines were in use eighty years before there was any scientific understanding of how they worked: Nicolas Carnot's *Réflexions sur la puissance motrice du feu* (about thermodynamics) was the product of and not the reason for the success of "fire-driven pumps."[11]

Today, science and technology have become so enmeshed in one another and so mutually reinforcing in their approach, functioning, and even results that it is better to talk in terms of a "technico-scientific system." Technical change now depends on both the laboratory and the factory, on science and manufacturing, on the university and the corporation. It was not until the nineteenth century, with the introduction of electricity and industrial chemistry, that the separate development of science and technology was replaced by interdependence—deliberate and organized—and that the synergy promoting ever more rapid scientific and technological progress emerged. It took nothing less than World War II to force governments to recognize science as a national resource on which depended not only their security but their chances of raising productivity and improving their competitive economic position.

Precisely what, then, is the share of truly scientific research—of basic research—in the undeniable and substantial influence of the technical system on economic growth? All answers to this question are inevitably subjective, because one cannot separate science from technology, and it is even less feasible to measure, in the famous "residual factor" dear to economists who specialize in research, development, and innovation, the effects attributable to the one independently of the other.

Nevertheless, it is the existence of the technico-scientific system, with its close interdependence, that obliges the decisionmakers in the most industrialized as well as the developing countries to ponder this question: Is it possible to ensure mastery of certain parts of the system without having to deal with it in its entirety? Would it be possible, for example, to get a handle on the processes involved in creating new forms of know-how without having to be involved in, or to acquire the capacity to contribute to, the main breakthroughs in the theoretical paradigm?

In other words, is it feasible to aim for a capacity for productive applied research—able to make use of existing knowledge to meet specific technical needs, perhaps even to develop this knowledge locally for practical purposes—without thereby having to take part in the business of explanation, which is the domain of basic research? If it is necessary to be involved in basic research, to what extent? In all the most advanced areas of research, or only a few? If so, which and at what cost—at the expense of what other commitments? These are questions no government today can evade, even if they are not raised as urgently or as shrilly in the developing and the industrialized nations.

Increasing Complexity

The early phases of the Industrial Revolution, which relied heavily on the links between technology and matter/energy, led to rapid growth in consumption of both. The current phase, characterized above all by the expansion of information technologies, shows a massive increase in the volume of information being handled. This does not imply that the technologies that use matter and energy are stagnating; on the contrary, they are undergoing rapid qualitative changes, though without any substantial growth in the quantities involved. Information is now the growing input, the element required for products and services to develop. This trend can be seen in manufacturing, where products are becoming more complicated rather than increasing in bulk.

Unlike operations involving matter and energy, this growth is unhindered by global physical barriers, in terms of the exhaustion of nonrenewable resources or disturbance of the ecosphere. The generic technologies underpinning this growth are still far from achieving the full potential that might be expected of them in theory. Meanwhile, needs continue to grow, sustained by forces within the process of social and economic development that seem unlikely to falter. For all these reasons, we may safely predict that both the absolute and relative expansion of information transactions are likely to continue. Doubtless, reliance on certain limited resources such as the electromagnetic field or the geostationary orbit will act as a constraint, but this involves *making use* of these resources, not *using them up*. This peculiar feature of our own day is in no way threatened over the long term.

Any technical activity results from the combination of a need—collective or individual—and a means to satisfy it. Collective needs always arise out of individual needs, so that technical activity ultimately deals with the basic needs of individuals. Let us put aside the question of which comes first, the need or the means, and concentrate on the essential difference between the two types of need.

The activities relating to matter and energy are usually quite directly linked to the basic human needs of the individual as an animal: food, clothing, shelter, defense, and good health. From the collective standpoint, the corresponding needs arise first from adding together these individual needs; the volume is therefore in proportion to the number of individuals and the opportunities for individual consumption, which can be infinite. Man obviously does not live by bread alone, and those human needs not strictly related to survival as a living being cannot be measured simply in terms of this arithmetic relationship. Furthermore, needs can be imaginary as well as real, and the whole notion of "need" is open to distortion—for instance, the effort to create weapons whose deadliness makes them unusable, or the economists' category of "conspicuous consumption." Nonetheless, if we limit ourselves to the need for food and for self-sufficiency in food, which is a vital matter for all developing countries, the quantitative estimate must be based mainly on population size.

The same is far from true of information transactions. The need for information arises directly and almost exclusively from the existence of social groupings. The human animal taken alone has few requirements: to mark one's territory and to recognize one's sexual partner. For all animals, social organization is accompanied by development of a language and a capacity to transfer from one individual to another the information needed for group cohesion and survival. In lower forms of life, this function is taken over almost entirely by genetic memory and instinct, whereas in human beings, the contents of memory—like language—is learned.

Unlike the consumption of food, the volume of information transactions required to maintain a social and economic structure is not directly related to the number of individuals concerned. The order of magnitude is determined by the degree of specialization of individual tasks required to maintain social and economic equilibrium for the group or society—put another way, by the complexity of the collective arrangements for seeing to the execution of these tasks.

For example, a community based on traditional farming and craft activities, managed by family units, will have little need to exchange information, and such information transactions as there are will be slow and over short distances. The same number of individuals grouped into a major technico-economic system (bank, airline, power distribution company) could not survive without a substantial flow of information over great distances and within ever shorter time periods, approaching "real time."

The size of the institutions now responsible for the running of industrialized countries, the number and degree of specialization of the individuals making up these institutions, plus the speed of their transactions in time and space are at once the cause and effect of the accelerating growth of information transactions. It is true these changes are occurring in parallel with an extension of systems and networks that are less prone to gigantism: Microelectronics makes it possible to avoid the tropism of megamachines and centralization. But everywhere the technico-economic system is expressed in an increasing degree of complexity that cannot be defined in quantitative terms.

It is obvious that the division of labor and the degree of specialization of individuals are one essential ingredient, but none of the top experts on the industrialized countries—economists, historians, or sociologists—has yet been able to come up with a satisfactory way of measuring this complexity. Measuring the flows of information is perhaps the direction to pursue, but as a yardstick it is both biased and incomplete, as if the total number of telephone calls were an adequate reflection of all the interactions within a society. This increase in complexity merely adds to the problems and lags facing the developing countries, most of which can master neither the tools nor the outcomes.

The Human Brain "Off-Line"

The information flows circulating within a society cater to a range of needs. They bring individuals the store of knowledge required to be a member of society and to hold employment; they ensure the maintenance of a community's political and cultural identity; they enable the productive system to operate and services to be provided.

The average stock of individual knowledge, the diversity of that knowledge found among individuals, and the way it changes over time are indicators of the level of development. This stock of information is what allows an industrialized country to return rapidly to normal after its productive system has been destroyed, as in the case of Germany or Japan after World War II. The volume of transactions necessary to create and update this stock of information (basic education, specialization, further education) increases as the level of development rises, and with it the extent and diversity of individuals' education. Growth of this stock is ultimately limited by the capacities of the human brain to learn and to store information.

The same holds true for the flows of information relating to the community's identity. Artistic events, political debates, news coverage, and sports and games have increased substantially with the introduction of modern technologies such as television. Yet this growth too, despite its size, is ultimately limited by the ability of the human brain to absorb information and the amount of time that individuals can devote to doing so. In the case of individual education, as for interactions at the level of society, the brain must by definition be directly engaged, or "on-line," to use the language of computers. It is impossible to conceive of a cultural activity or game involving just machines, from which human beings were excluded.

Matters are quite different when it comes to the productive system.

There is no obstacle to the ability of machines to replace the human brain in executing tasks needed for the productive system to function, so that there is no reason ever to reach saturation point because of the limits of the human brain. On the contrary, the development of the contemporary productive system shows that the scale of the systems that cater to all the various economic needs, just like the volume of information transactions required to sustain these systems, is growing all the time. Both have long ago outstripped what human beings are able to manipulate and absorb without help from machines.

The peculiar feature of the contemporary technico-economic system is its tendency to eliminate the human brain from the productive circuit. Indeed, it is this shift to putting the human brain "off-line" that is at the basis of all the promises and threats associated with automation and robotization. And there seems to be no limit in theory to this process of substitution. The work of John von Neumann revealed the possibility of robots capable of reproducing themselves: Technical artifacts can acquire the capacity not only to keep themselves in running order but also to increase autonomously.[12] We are just in the earliest stages of this process of substitution. It may be impossible to foresee where the process will end, but there is no doubt about the trend. It moves from the most readily predictable tasks, which are repetitive and can be expressed in an algorithm, toward tasks that are closer to those traditionally attributed to human intelligence, requiring judgment and versatility.[13]

Already many technical plants operate without human intervention once they have been assembled (e.g., oil refineries, nuclear power stations). Some human presence is nevertheless still required to keep watch and deal with technical faults. When these enormous complexes go wrong, the risks are such that only human intervention can manage the course of events.[14] Thus, the contemporary technical system may be relying on ever increasing numbers of robots, but the robots still need human beings. And the economic and social conditions must also be favorable if the robots are to go on increasing.

As Paul David has shown, robots will be used on a large scale only if there is a significant fall in the costs of installing, maintaining, and repairing them. At the moment, only the biggest firms can afford the costs of automation, which do not involve merely the purchase of the machines but also the modifications to the manufacturing process entailed by the new machines, not to mention the problems of adapting the workforce. David compares the future of robots to the experience with combine-harvesters: First available in the United States around 1830, they were not widely used until about thirty years later—and this occurred not because of the technical improvements made but because of the international boom in wheat during the 1850s, which started to bring down the price of the machines relative to farm wages in the Midwest.[15]

For some factories, automation is already a factor in their competitive position, but the manufacturers of robots and the media have been rather too quick in announcing that there is widespread use. Large-scale use of robots faces the same obstacles the combine-harvesters encountered: the cost of the machines, the attitudes of firms, and the macroeconomic situation. The market was still so far from being ready in 1985 that another U.S. observer joked that the manufacturers of robots had not made large profits, and indeed many were having a hard time finding jobs for these tireless—and mainly unemployed—workers.[16] But what of the future? The trend toward automation seems as inevitable as was mechanization after the first successful industrial applications.

From this standpoint, if there are reasons to be anxious about the outlook for employment in the industrialized countries, there is no cause at all to be optimistic about the developing countries—all the less so in that the contemporary technical system is only very remotely concerned with the problems of developing countries. It is only in indirect ways, via difficult transfer procedures and often via its byproducts, that the modern system offers solutions to their specific needs. The overabundance of information flows that now fuel the industrialized world can only lead to congestion and a destruction of cultural identity. In the future, above all, when the robots in the industrialized countries are no longer unemployed, it is hard to see what jobs will be reserved for the huge mass of workers in the Third Worlds.

5 The Science of the Poor

> Three decades of development in Asia, two in Africa, a significantly longer period in Latin America offer an ample range of experiences that ought to enable us to put together development theories based on facts and not just speculations or references to the paths followed by the countries which are now industrialized. And yet dogmatism and rhetoric flourish as much as ever.[1]

When the topic is "science and technology in the service of development," people may well talk about science, and the speakers may well often be scientists, but the discussion is no less prone to fall into the traps of demagogy, as was clear from the conference on this subject organized by the United Nations in Vienna in 1979.

More than two thousand participants, tons of documents, and dozens of speeches by ministers, ambassadors, and experts revealed as many points of discord between North and South as within the Group of 77 (which in fact number 117), representing the developing countries. And the results were insignificant. The South expected the North to make concessions in the form of gifts, lower prices, and licenses and patents offered virtually for nothing as forms of assistance in technology transfers. In the years leading up to the conference, technology had been perceived by the Third Worlds as just another commodity like any other; the poorest countries' demands to be given access to the technologies owned by the industrialized countries came down to wanting the costs cut, as was presented in the negotiations of the United Nations Conference on Trade and Development (UNCTAD).

The only positive outcome of the conference was to show that technology cannot be treated like merchandise, which would be traded more freely if the tariff barriers were only lowered. As was stressed by the secretary-general of the conference, Brazilian ambassador João da Costa, technology should be seen as the product of a particular economic, social, and political system. We should therefore distinguish between the problems related to the manner of its acquisition (licensing, monopolies, restrictive practices) and those that arise from a more subtle type of dependency, inherent in the nature of the technologies being transferred and therefore in the business of mastering them. He went on to say that factors in the second category impose foreign standards, forms of organization, and cultural values, which encourage neither assimilation by the recipients nor expansion of scientific research specifically geared to the needs of the Third World.

There were plans to create a special UN agency responsible for coordinating research programs designed in accordance with these needs, but the Western nations were no more enthusiastic than the communist bloc about supporting this idea. In the corridors of the meeting, the delegates on all sides made no bones about it: Why create yet another agency, costing millions of dollars, run by international bureaucrats? The conference gave birth to a midget: the Office of Science and Technology for Development, with a small secretariat and a minimal budget, condemned to conduct studies and organize seminars at UN headquarters in New York.

The failure of the Vienna meeting cannot be explained entirely by the reluctance of the industrialized countries to countenance yet another agency with an unclear mission, nor by their economic problems (they were caught in the stranglehold of rising oil prices and ever slower growth rates). In fact, lacking any commonly agreed definitions of what was meant by development or (more important) of what might be hoped from scientific research and technology transfers, the meeting could hardly have been other than a dialogue of the deaf. Despite speeches of solidarity and joint motions, the developing countries clearly could not speak about these problems with one voice: Science is universal, of course, but the distribution of scientific vocations, competencies, and resources is not. At a deeper level, what does the "universality of science" really mean, beyond the structures and the institutions that the scientific method is based upon?

When Universal Is Not Universal

The scientists at the conference who were members of Third World delegations could appear in sympathy with the rest of their colleagues from the South. They were on equally good terms, thanks to education and intellectual common ground, with their colleagues from the North, and indeed often it was easier for scientists from the North and South to talk to one another than it was for scientists, diplomats, and politicians from the South to see eye to eye. The real divide was not between the rich and the poor nations, or even between the capitalists and the communists, but between those countries, rich or poor, having a history and culture that had given them a long preparation for modern science—the experimental method but also the worldview that it entails—and all the others.

The international network of scientists trained in the same institutions of higher learning and research, speaking the same language, publishing in the same journals, and meeting each other from time to time at symposiums in the same places is indeed based on a commonality of language, methods, and products of a universal scientific community as understood in the West. A phenomenon or a proposition is said to be universal if it applies to the whole world, but—like airline routes—the network of international science is not universal in the sense that *everybody has equal access*. Indian scientists share in the universality of science because of having been trained in the procedures that are the basis of Western science; consequently, they feel closer to their colleagues at King's College or the Sorbonne or MIT than they do to the villagers of their own land. Most of them remain excluded from the network, however; universality stops there.[2]

The existence of the international scientific community means simply that if laboratories were extended like airline routes, scientific ideas could be moved about like planes. But planes do not land just anywhere—they need an infrastructure, which involves not only physical runways but also highly trained staff responsible for maintenance and air traffic control. One international airport looks much like another because they have similar functions, as well as because their architects adopt similar styles (to such a degree that their hotels give one the impression of being anywhere and nowhere). But as soon as travelers leave the airport for the city proper, they are reabsorbed into the characteristic features of the national culture, shaped in the first place by climate and geography.

This does not mean that European or North American scientists, once outside the laboratory, never meet people who do not remotely share the cultural values of the scientific community (nor that scientists never read their horoscopes—the success of astronomy has not undermined interest in astrology). But the social context in which they live is deeply affected by the logic of scientific method and organization. In so-called advanced countries, although superstition and belief in magic have never been completely repressed, they are nevertheless not a general point of reference or a guiding principle in people's lives. By contrast, in many developing countries, laboratories are inserted into environments that are imbued with the exact opposite of the scientific approach.

Fortunately, nobody now makes invidious comparisons between the reasoning of "the white man, mature and civilized" and that of "the primitive mind," which was thought to be "a rudimentary version of our own, childlike and almost pathological."[3] Anthropology, ethnology, and comparative sociology have revealed the consistency and legitimacy of ideas and behavior that are not the product of Western-style rationality. This other type is always to be found in industrialized societies, but it tends to be marginal and to disappear as education based on scientific ideas and methods becomes more widespread. In the developing countries, allowance must be made for it in mounting efforts to spread the technological culture evolved in the industrialized countries, to train researchers, and to gain recognition for institutions that purvey and improve upon scientific knowledge.

> The question: "Are there cultures which are more or less receptive to Western (European) science," [nevertheless] elicits a positive response. If the sources of this science are traced back to the mathematization of nature and to experimental proof, it is clear that any culture professing an order of natural phenomena in which mathematical rationality is not essential is hardly made to adopt the Western approach. This response does not signify that Western rationality challenges this culture as such; it simply defines it as *other,* and the difference of its practices does not at all mean that *they are not operational.* But this otherness does not claim to have a power of universal application there where Western rationality is only so when postulating an order—the constancy of the laws of the universe—of which mathematization and experimental proof take advantage to act universally on natural phenomena.[4]

From Lucretius, who spoke of the laws of nature as contracts, to Einstein, who proclaimed that "God is subtle, but does not have a malicious nature," the postulate of this rationality is that the universe functions according to commands that are like inalterable and universal decrees. In fact these would seem to be the decrees of a suprarational lawgiver, decrees the founders of modern science—Galileo, Descartes, Kepler, Newton—thought to be "revealed" to the human spirit. This is what led Joseph Needham to demonstrate quite concretely the essential difference between the conception of the order of the world in traditional China and that in Renaissance Europe. In the latter, the laws of nature are valid for heaven and earth according to orders given by a rational lawgiver; in the former, there is no superior authority instituting a system of causal relations but an organic cooperation defining a cosmic reality: The law has no clear representation outside human affairs so that the intelligibility of the world is never guaranteed.

Needham cited the example of medieval Europe, struggling against sorcery, where trials were held in which charges were brought against roosters that had laid eggs. These roosters were condemned to be burned alive because they had betrayed the divine order. Taoist China would never have dreamed of conducting similar trials. Such phenomena were considered to be "reprimands from heaven," "caerulean misfortunes," and not a perversion of the order of the world guaranteed by God. Western science was finally developed and imposed itself, after the seventeenth century and above all in the eighteenth, by doing without the guarantee of a supreme lawgiver. Nevertheless, statistical regularities and their mathematical expressions are still guaranteed by the hypothesis of an always-honored contract, of an order removed from the whims and arbitrary moods of either a magical or a malicious intervention—hence the remark by Needham that marvelously locates the boundary between the cultures ready to adopt a Western rationality and those closed to it: "Was perhaps the state of mind in which an egg-laying cock could be prosecuted at law necessary in a culture which should later have the property of producing a Kepler?"[5]

Different rationalities are not merely to be perceived as relics of the past surviving in industrialized countries: They are thriving even in the midst of the scientific establishment, as is clear from the way that the teaching of Chinese acupuncture has spread in Western medical schools, or from the return to herbal remedies and "soft" technologies. There is thus all the more reason for traditional and modern methods to continue to exist side by side in developing countries, complementing one another. For the health services of developing countries, this complementarity in fact accords with social necessity. In Asia, the popular medicine provided by herbalists, soothsayers, spirit mediums, and Taoist or Buddhist priests carries on alongside scholarly traditional medicine practiced by people trained in recognized schools and hospitals.[6] The usefulness of the two together is all the greater in that Western medicine is costly, beyond the reach of the majority, and impossible to provide in rural areas. The one benefit of the Cultural Revolution in China—in spite of the massacres, deportations, and chaos—was probably to increase the numbers of "barefoot doctors" in the country areas; without them, the efforts to improve health education and preventive medicine would not have been so successful.

The two different styles of medicine can remain on good terms, in much the same way that religious syncretism unites the gods of the past with the new system of belief. In China and India today, traditional methods are often combined with modern chemotherapy techniques. In India, much research is being conducted on the results obtained by ayurvedic medicine based on plants; in Japan, Hong Kong, and Taiwan, publications about the traditional medicine called *kanpo* have enjoyed a tremendous boom, and it is not uncommon for a Japanese doctor trained in Western methods to practice *kanpo* at the same time.

But once we pass beyond health matters to the domain of scientific research—where what is at issue is not merely to heal the sick but to contribute to the advancement of science—the relationship between the two rationalities is less cordial. Doctors are not particularly bothered about whether their skills derive from traditional or standard Western medicine; they know they are useful to the societies where they live. On the other hand, scientists working in high-energy particle physics, solid-state physics, or molecular biology may well wonder where their field ranks in the order of urgent priorities or human values in their countries. The antagonism between the rationalities can become so sharp (and can be so vividly experienced by researchers) that they rail against the deep gulf between the methods based on Western rationality and those of the culture where this rationality often seems like an intruder.

"Science has grown as an oasis in an environment which, if not antagonistic, is also not sympathetic to it, with the majority of people steeped in traditions and traditionalism of which many of the leading scientists are also victims."[7] What Abdul Rahman, head of the National Institute for Research on Science and Development in New Delhi, wrote of India in the aftermath of independence is equally applicable to many other nations. Nowhere more than in India has the concern—indeed the critical need—to combat these superstitions seemed more pressing to both scientists and politicians. As soon as independence was declared, Jawaharlal Nehru issued the command to teach science so as to promote the "scientific temper," not so much in providing instruction in scientific knowledge as encouraging the spread of attitudes based on scientific rationality. This crusade was carried on relentlessly by his successors (his daughter Indira and his grandson Rajiv), and many scientists continue to devote part of their time to it by educating the general public.[8]

Investment in Research

In a special report published two decades ago in the magazine *La Recherche* under the title "The Science of the Poor," Giovanni Rossi listed a series of obstacles that together go some way to explaining why, despite two "decades of development," the contributions of science and technology to the fight against underdevelopment remain limited.[9] Since the 1970s, some Third World countries have made substantial efforts to build up their resources in science and technology (S&T), while international agencies have done much to help them formulate and implement sound S&T policies. Nevertheless, the successes can be counted on the fingers of one hand, and they are in any case relative in comparison with the challenges of development. If the findings in Rossi's article are brought up-to-date, and the achievements of the newly industrialized countries are added as well, the overall conclusions remain much the same as before.

The race involving science and technology is not only very unequal, but the handicaps seem set so as to favor the strongest runners. Nothing hinders the most competitive nations, whereas in the countries with the least productive investments, the social costs incurred are also very high. In order to keep or lure back their own scientists, underdeveloped countries have to offer far higher salaries and more attractive living conditions than the rest of the population enjoys, and the equipment has to be imported at prices dictated by costs in the industrialized countries.

We have seen how unevenly distributed R&D efforts are in the industrialized countries, and the gap between them and those of the Third Worlds are perforce greater still. Prior to 1989, the developed economies (including the communist ones) accounted for 94 percent of R&D expenditures and 89 percent of the scientists, engineers, and technicians, leaving the Third Worlds with just 6 percent of the expenditures and 11 percent of the researchers. One has only to glance at Figures 5.1 and 5.2 to appreciate the enormity of the disparities. In this area, the dependency of the poor is truly the obverse of the domination of the rich.

The scientific and technological activities of the industrialized countries give rise to techniques, products, and processes that are geared mainly to satisfying their own market needs and only rarely those of the developing countries—where they are often useless, if not actually harmful. Internationally, roughly 85 percent of patents are registered by OECD member countries and only about 6 percent by developing countries. Not only is the research conducted in the developing countries marginal in comparison with the work being done in the industrialized countries, but it does not even attempt to provide solutions to local problems of health, food, energy, or the urban or rural environment.

The Third Worlds' position in science and technology is by definition precarious, sometimes pitiful. The data available on R&D are inadequate indicators: Official statistics (suggesting spending equivalent to 0.1-0.5 percent of GDP) are often inflated and are not based on standard definitions when they are collected. Furthermore, the figures do not take account of the differences in the way research is organized. There are a few research centers that rival those of the top Western institutions, but in most cases research activities are scattered across a large number of small laboratories or centres, which are underfinanced and underequipped. The quality of the research thus is lowered even further, whether undertaken in universities or government establishments.

On the whole, biomedical and agricultural research geared to the specific problems of developing countries is carried out in the industrialized countries. For over a decade, India spent 40 percent of its research budget on nuclear physics and only 8 percent on agriculture. The results obtained, especially in military terms, prove that neither resources nor brains were lacking, but the priority given to defense automatically meant less for everything else.

Imbalance of this kind is not just the result of developing countries' own preferences; it is also partly a reflection of the research topics encouraged by the agencies in the industrialized countries that provide financial support for research teams in the Third World. The share of research in developing countries financed from abroad has risen steadily since the

Figure 5.1 Distribution of R&D Scientists and Engineers and Expenditures (1980 estimates)

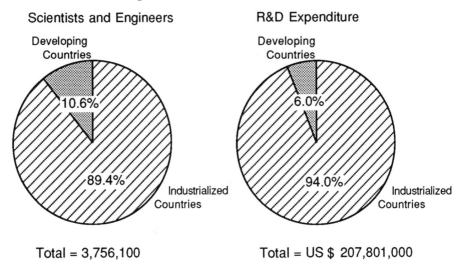

Figure 5.2 Distribution of R&D Scientists and Engineers and Expenditure by Major Areas (1980 estimates)

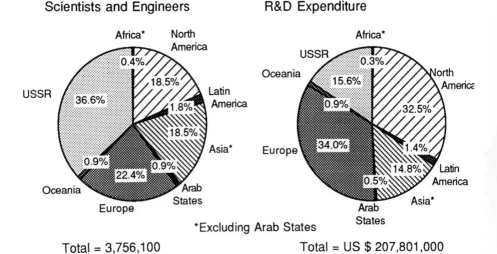

Source: Data for the above figures are from UNESCO, *Statistical Yearbook 1983*, Paris: UNESCO, 1983.

1970s—in some cases, for example Senegal, the figure is as high as 70 percent or more—which indicates the influence the West exerts on the choice of research topics in the Third World.[10] The publicity given to these topics, for which it is possible and tempting to obtain funding, can encourage scientists to undertake research into problems that are far removed from urgent local needs.

The attraction exerted by international research contracts can reach absurd proportions. Because of a U.S. report that was distributed widely in the developing countries, there has been a craze recently for studying a variety of bean (*psophocarpus tetragonolobus*), a tropical legume rich in protein that has the advantage of being 100 percent edible (including the roots, leaves, and stalk). Research teams threw themselves into studies of this vegetable in many Third World countries, even though the climatic conditions and the normal diet of the people more often than not ruled out its consumption locally. And the "miracle food" did not fulfill the promises too precipitously suggested by the international publicity.[11]

The Magnet of Pure Science

Modern research may bring science and technology closer and closer together, but the motives of researchers are not the same when they are working on fundamental, theoretical problems as when they are tackling applied problems, especially in industry. Basic research, normally carried out in universities, is an open system relying upon discussion and recognition by specialists of results published in scholarly journals. The international scientific community is defined by a network of contacts and institutions that function as a channel of communication and regulation: The scientists offer the results of their efforts without charge to their colleagues throughout the world; in return, the work is assessed, recognized, and sometimes rewarded by specialists in the same field.

The regulatory role is filled by journals, referees, peer review committees, and learned societies and Academies—the visible and invisible network of "equals." The rewards range from funding for research projects to promotion up the career ladder and scientific prizes, the most prestigious being the Nobel Prizes (given annually in physics, chemistry, biology, and economics) and the Field Medal (awarded once every four years in mathematics). International recognition for a scientist is not unlike canonization by the Vatican, which has commissions constantly investigating the suitability of candidates for sainthood, the main difference being that the scientists are still alive.

Applied research, by contrast, operates under very different systems of assessment and reward: Results are patented or licensed and are shrouded in industrial secrecy. This is what caused Derek de Solla Price, when seeking to distinguish science from technology, to say, "The scientist wants to write and not to read, and the technician wants to read and not to write."[12] The fact is that basic research cannot exist without publication and open sharing of results, because criticism in itself both guarantees and contributes to scientific progress. The system of reference is inevitably transnational, quite outside parochial worries and contingencies, unconcerned with practical needs in general and with the things developing countries most urgently require in particular.

Basic research engages Third World scientists in work on subjects that are being investigated by the best brains from the best universities in the industrialized countries. The interest of these topics lies not in the technical problems that must be overcome to meet the economic or other needs of a particular country but in the theoretical and practical questions faced by the various disciplines in order to extend the frontiers of knowledge—truly a view from the ivory tower, apparently far removed from the everyday problems of society. Basic research as part of the realm of pure scientific speculation goes back to the very beginnings of Western science, with the separation of theory from practice first made by the Greeks. In antiquity, the two distinct types of knowledge were also connected with a social division: Practical skills were left to the lower, "servile" classes, whereas theory was the preserve of the upper, "free" classes.

Although experimental science developed from the seventeenth century onward by breaking with this approach, some traces of it nevertheless remain in the universities. There is still a notion of "pure (basic) science," regarded as more attractive, prestigious, and noble than applied research that requires scientists to "get their hands dirty." The image of the scholar as a figure of culture contrasts with that of the scientist who works like everyone else: The former "thinks" science in the cultural context of the world of ideas, whereas the latter "does" science within the productive system.[13] An Indian scientist has explained the success enjoyed by Western science and astronomy in the nineteenth century among the elite of Bengal (Brahmans, Baidyas, and Kayaths) not simply as a way of making connections with the Hindu tradition of contemplative knowledge but also as a way of differentiating themselves from the Muslim lower classes who were then engaged only in practical trades.[14] Even in the industrialized countries, however much the evolution of research brings the two worlds closer together, the separation is lasting, at least in the social perception of science.

There is in any case nothing remarkable about the fact that basic researchers in the Third World behave as if they were closer to their peers than to their fellow citizens. Their training, their intellectual interests, and the links made in the course of their studies with scientists in industrialized countries all explain this preference for pure science just as much as does the security of the university environment or the absence of research opportunities provided by local industry. The international attraction of basic science arises not merely because of the widespread influence of research scientists but also because scientists share a language, whatever their nationalities or ideological differences. "Newton is my colleague, and Galileo," said the Nobel laureate Albert Szent-Györgi, adding so as to emphasize that the scholarly community transcends all frontiers: "A Chinese scientist is much closer to me than my own milkman."[15]

The selection, training, and values of the basic researcher are polarized in such a way that professional demands tend to be distinct from personal social commitment. In pure science, as in literature, the right feelings are not enough to ensure a masterpiece; rather, there is a risk of producing substandard science. Szent-Györgi also said that if a student came to see him, expressing a desire to help humanity and do research in order to relieve human suffering, his advice was to work for a charity; science needs really selfish people who do research for their own pleasure and satisfaction, which they find in solving the riddles of nature.[16]

The pleasure of searching—and sometimes finding—may be an aspect of the thirst for knowledge (innate in human beings, according to Aristotle), but the basic researcher's motives obviously cannot be explained purely in terms of hedonism. Yet it is impossible to deny this peculiar feature of basic science: Its contributions to the welfare of the human race occur via extremely tortuous routes leading from research that is not aiming in the first instance to solve social problems. Hence arises the contradiction often remarked between the vocation of this type of research and the aspirations of some scientists—and, more generally, of certain political parties or governments—to put science "to work for society." The pursuit of knowledge, as understood and encouraged by the international scientific community, may indeed be in the optimistic and liberating tradition of the Enlightenment, as is claimed, but its most obvious purpose is not that of satisfying the demands of society.

Let us go further and put the question bluntly: When a developing country does manage to conduct high-quality basic research, which is internationally recognized and on a par with that of the leading industrialized countries, what impact does this have on its level of development? The answer has to be—even if this is shocking—*none whatever*. Basic research is not needed in order to develop efficient technical and industrial activities. We have stressed all along the close reciprocal links (if not dependence) that the evolution of contemporary research has brought about between the scientific paradigm and the technical system. But it does not follow that one must contribute to the progress of the paradigm in order to be sure of mastering some part of the technical system.

Basic research helps in developing skills that in turn make it possible to understand how the technical system works and therefore to benefit from it; such research is not essential just to use the technology in question.

It is indeed indispensable in order to communicate the latest developments to those who are themselves in the avant-garde of scientific discovery; it is not needed to teach scientific principles to people who will simply be applying them in their own work.

The ingenuity of the Japanese is legendary in both traditional and very advanced technologies (electronics and biotechnologies), but this reputation owes little to basic research or even to Japanese universities. The staggering technical and industrial success of Japan has so far involved relatively minor contributions to scientific progress as such. If this situation is beginning to change, it is doubtless because the increasing complexity of the technical system, which has hitherto been mastered via technical imitation rather than scientific creation, now requires a much larger input of theoretical research. But this is also because the economic prosperity of the country, achieved as a result of the Japanese mastery of the technical system, has allowed this change to occur, rather than that it has been forced upon them.

The Bird with Clipped Wings

Scientists in developing countries face a very different situation: With their gaze directed toward the research topics and the values that drive the best laboratories in the industrialized countries, they are often condemned to working in conditions quite unlike those of their role models. To stay in the forefront of scientific research requires substantial funds. The purchase of books means spending precious hard currency; equipment costs far more; the scientific journals must be delivered quickly and therefore must come by air at great expense. In addition, because most developing countries are tropical, the equipment must be able to withstand heat, humidity, mold, and insects. For the laboratories, the extra costs are enormous to maintain instruments and machines, such as air conditioning and refrigeration for computers.

Furthermore, the customs regulations in most countries are inappropriate, preventing the importation of essential equipment and holding up the arrival of spare parts or chemical products needed for experiments. The telephone systems work badly, limiting the informal contacts between researchers in different countries that are sometimes more crucial to the success of a program than encounters at formal conferences. Administrative and political arrangements impose bureaucratic regulations and often insist on secrecy, even for work that has no bearing on national security.

Michael Moravcsik, a U.S. physicist who spent part of his career working in Third World countries, has compared the scientist in developing countries to a bird whose wings have been clipped but nevertheless tries to fly.[17] Flying has become increasingly difficult since the end of the

1970s. The numbers of science graduates in the Third World, more and more of them trained to predoctoral level in local universities, have risen substantially. But working conditions have not improved and have even deteriorated in many countries (e.g., Argentina, Chile, Mexico, Peru, Kenya).

Rising energy prices, galloping inflation, and debt have added to the problems of the best research institutions, even in countries like Brazil and India that have a well-established tradition of research and that have put a great deal of effort and money into promoting science in the last two decades. There is nothing worse for a research system than "stop-go" support—sometimes there are not enough well-qualified scientists to carry out the programs, at other times there are too many in comparison with the funds available. Recent studies confirm both that conditions have worsened and that researchers are increasingly disenchanted, lacking both the stimulus and the stimulants enjoyed by their counterparts in the industrialized countries.

Science in developing countries was already marginal, and since the oil crises it has become even more so. Quantitative studies of the scientific literature based on the *Science Citation Index* (which measures world scientific output, discipline by discipline) highlight the paucity of Third World contributions to scientific progress. Articles from the South are rarely published in journals with an international readership and thus are rarely cited. This "scientometric" approach is admittedly biased in many ways (in particular because its coverage is almost exclusively English-language journals), but the orders of magnitude revealed are all too plausible. In short, the science of the periphery is on the periphery of science.[18]

Qualitative studies show much the same picture. An example is the comparative study undertaken by Thomas O. Eisemon; it was based on interviews with teachers and researchers in the mathematics and zoology departments of the universities of Ibadan and Nairobi.

> The achievements of Nigerian and Kenyan science are primarily quantitative and in the sphere of the construction of an institutional framework for scientific research. Science teaching programmes have been developed, scientific societies established, publishing institutions formed. These are no trivial accomplishments in my view. Nevertheless it is also true that scientific work—in a more substantial sense—has not been much advanced. Predictions of the weakening of the ties to international scientific work resulting from policies favouring Africanisation and the articulation of scientific activity with "national development" have not been fulfilled. Nor have hopes for rapid scientific development been realised. A much longer time will be required before a conclusive judgment can be passed on the effective implantation of the scientific ethos in Black Africa.[19]

India, a country of great contrasts, has laboratories doing leading-edge research in some fields as well as a host of poorly equipped universities. The five major national technological institutes (Delhi, Bombay, Kanpur, Kharagpur, Madras) and the handful of independent institutes of advanced research (the Tata Institute in Bombay, the Saha Institute in Calcutta, the Institute of Theoretical Physics in Madras) are home to the best research groups and receive priority support, but the labs on many campuses are in a parlous state. "Excellence in the Midst of Poverty" was the title of a special report on India published in 1984 by *Nature,* the most widely read scientific journal in the world.[20] One could in fact interpret the title in several ways, emphasizing either the high quality of certain institutions and research programs, the lack of finance and equipment in most Indian research institutions, or the vast problems of the country that only a tiny proportion of scientists attempt to solve.

There have indeed been several generations of top-rank Indian scientists, such as C. V. Raman (Nobel Prize for physics in 1930), S. N. Bose, M. N. Saha, H. J. Babbha, or M. S. Krishnan, most of them trained in the best laboratories in the United States and Britain. They made major contributions to the advance of science. But they are like seaspray, covering a range of institutions of varied quality. In some areas (mathematics, astrophysics, or more recently, oceanography), Indian science has an excellent reputation; in others, as a 1980 survey of various branches of physics showed (solid-state physics, materials science, metallurgy, and particle physics), the research and the publications are well below international standards.[21]

The case of Indian science shows that it is not enough to follow Western models to achieve the same working conditions for scientists; researchers seem themselves to be marginal within Indian society. The language of science is, naturally, English—spoken fluently by 3 percent of the population—while there are fifteen official languages and, above all, more than 750 million people who are illiterate. "Excellence in the Midst of Poverty" also brings out the size of the gap between the institutionalized forms of research work (modeled on the best institutions of Europe and North America and in some cases on a par with them) and the daily reality in which this work is done, work that very often entails struggling with all the difficulties of underdevelopment.

A study of science in the Arab world suggests that the achievements of policies for training scientists and technicians should be treated with similar caution. The author, A. B. Zahlan, a physicist from Beirut who has become a historian of science and one of the top experts on the problems of science policy in the Arab world, shows that the numbers of university graduates has risen sharply over the last thirty years or so, virtually doubling every five years: In science, there were 760,000 in 1975 and 1.5 million in 1980. If this rate were maintained, there would be almost 12 million altogether (in all subjects) in 2000.

However, the variations among countries are so great that aggregating statistics in this way is misleading. It is nonsense to count up the scientists in the Arab world as if they were homogeneous, when in fact they come from countries as different as Algeria, Egypt, Iran, Iraq, Jordan, Kuwait, Lebanon, Libya, Morocco, Saudi Arabia, Tunisia, and Yemen. If we assume for the moment that to add them together made some sense, Zahlan shows that the numbers are out of all proportion to the research work undertaken and the significance of the results, because the political, social, and cultural factors in most of these countries are a hindrance to the expansion of a productive research system:

> When national universities are established, they are often shallow reflections of the institutions being copied. Institutions are set up without those conditions most essential for intellectual and scientific activity: a faculty that is socially, economically and intellectually secure; library and laboratory facilities; academic standards for both faculty and students; and tolerance for the opinions, findings and criticisms emanating from research and academic institutions.
>
> The absence of such conditions in most of the Arab countries is reflected in both the massive brain drain (probably approaching 50 percent of Arab doctorate holders) and the extremely poor productivity of scientific manpower employed in the Arab world. Arab scientific manpower in 1979 is numerically comparable to that of the US, and exceeds that of the UK or Japan during World War II. Yet if one were to compare the performance of these different groups, albeit at different times and under different circumstances, one would be hard put to find any scientific achievements in the Arab world comparable with the output of the other scientific communities. Obviously the causes of these differences lie in the state of the institutions rather than in the size of the degree-qualified manpower.[22]

The following analysis by another scientist reveals exactly why scarcity is not the only or perhaps even the most important obstacle to be surmounted if the research systems of the Third Worlds are to become as efficient, if not as creative, as Western ones:

> It has been pointed out again and again, by many authors, that it is immensely difficult to organize research institutes in underdeveloped countries and to keep abreast with latest developments. However, the point I want to make here is that, if we tacitly assume that these are the only obstacles in the way of starting a new scientific tradition, and that science would start blossoming in an underdeveloped country as soon as efficiently organized institutions are set up with good libraries and adequate research facilities, then we would be missing a most vital aspect of the problem. Science is one of the profoundest forms of creative expression of the human mind. Unless we have human minds properly conditioned to create science, it is absurd to expect science to stream out of buildings, libraries and laboratories, however well-equipped they may be.[23]

All in all, in the Third World scientists trained to do basic research are faced with three sources of alienation: They must cope with conflicting cultures; socially they are outsiders, seeking results that are remote and abstract, and completely obscure to the vast majority of their compatriots; and compared with their counterparts in the industrialized countries, they often feel distanced, if not actually excluded, from the most exciting avenues of research.

The Brain Drain

The overproduction of scientists, educated according to Western standards and in research areas that interest industrialized countries, contributes directly to the outflow of researchers at the same time that there are too few institutions in the developing countries in a position to train technicians and middle managers, they are ill-suited to their task, and they do not attract the most gifted students.

The phrase "brain drain" graphically captures the idea of the loss of scientific skills from the periphery to the center. Twenty years ago, when people still had a very hazy notion of the scale of the phenomenon, the brain drain generated passionate debate in international circles. Some developing countries demanded some kind of reparation for the losses they suffered and the savings the industrialized countries were able to realize by acquiring experts without having to pay for the early years of their education.

In fact the brain drain is much older than the twentieth century: Think of Plato offering his services to the tyrant of Syracuse, Galileo taking refuge in Venice, Descartes accepting the invitation of Queen Christina, the Cassini brothers attracted by Louis XIV's gold to build the Paris observatory, Huguenots fleeing the Counter-Reformation or aristocrats emigrating after the French Revolution, or the scientists recruited in Europe by the first major U.S. industrial laboratories (by Edison, especially to Menlo Park). There are as many reasons as examples, all quite different, for scientists choosing or being forced to emigrate—from the lure of travel, of novelty, of pecuniary or intellectual benefits to flight from persecution. The flow has increased considerably in the last forty years, partly because of the growth in international contacts, partly because science and technology play a decisive role in world affairs. The debate about the brain drain has, however, died down, because it is impossible to check the movement of scientists in the same way as controls are placed on movements of capital, goods, or prices.

The only way to check the brain drain is to forbid emigration, as totalitarian regimes have done. Talent by definition goes not where it is hoped it will go or stay but where it feels wanted. As far as scientific research is concerned—and intellectual and creative freedom more generally—the forces of attraction are indissolubly linked to those of repulsion. What draws people abroad is precisely whatever is lacking at home. All the studies of the brain drain agree in finding that the structures provided for research work are just as important, if not more so, than financial inducements.[24]

A famous French actress who was brought to trial after the war for having sexual relations with German officers retorted, when accused of having collaborated, "Your Honor, they should never have been allowed in!" The emigrant scientist could equally say, "You shouldn't have allowed me to leave!" The many reasons for leaving—or for staying abroad, in the case of specialists who have gone there for further training—differ among individuals, countries, and circumstances; examples include lack of flexibility in state-sector employment, institutional inertia, scarcity of public funding for research, professional isolation, the barriers to promotion, access to jobs through relatives or friends in power rather than merit, the difficulties of younger people seeking recognition of their talents, and prejudices based on race, tribal origin, or caste. Political circumstances have also had a major impact: In the past, they explained much of the emigration of European scientists, Jewish and others, to the United States during and after World War II (more than 65,000 between 1939 and 1960, among them a cluster of Nobel laureates); nowadays, arbitrary regimes, intolerance, and repression force many developing-country scientists into exile.

The gain for the receiving country is substantial, in line with the loss (even if it is impossible to calculate) to the Third World. Statistics do not accurately capture the full loss involved. Charles Kidd of the National Science Foundation estimated that the annual emigration of doctors (practitioners rather than researchers) from developing countries to Britain (several hundred) and the United States (about 2,500) is equivalent to the number of graduates of twenty-five new medical schools.[25] Yet these countries need physicians even more than researchers.

A U.S. government study reckoned that in 1973, the United States saved $883 million on the costs of education, while developing countries lost $320 million (the sum spent on educating the emigrants). Another study in Canada found that the cost of educating the equivalent of the human capital that settled in the country between 1967 and 1973 rose from $Can1 billion to $2.4 billion in 1968 prices, or ten times more than the amount Canada spent in the same period for improvements in education and technical assistance.[26]

Naturally, none of these calculations takes account of the loss of output the exporting country may have incurred through the departure of well-qualified technicians and managers, though it should also be recognized that this "opportunity cost" should be set against the true employment possibilities for scientists in Third World countries—in fact extremely limited. Thus, the brain drain may free some developing countries of specialists they cannot use rather than deprive them of professionals they really need.

It is characteristic of underdeveloped countries to produce more graduates than their resources and facilities can absorb but fewer middle-level managers and technicians than their economies require. If one looks at the most urgent requirements for doctors, technicians, managers, agronomists, and primary and secondary school teachers, it is clear there is an enormous shortage of qualified people. By contrast, if one looks at the numbers of university graduates who cannot find the jobs they would like, the situation is one of surplus rather than shortage.

India, for instance, has record numbers of emigrant scientists, and those who leave are probably among the most gifted. Moreover, India has one of the highest rates of unemployment among engineers and scientists in the world. Huge numbers of scientists and engineers have been trained there, rising from 180,000 in 1950 to almost 2.5 million in 1983. Professor Abdus Salam, the Pakistani Nobel laureate and director of the Institute of Theoretical Physics in Trieste, described in an interview how proud Indian scientists are of the fact that their country now has the third largest scientific community in the world, leading him to conclude that India would be a superpower by the year 2000.[27]

These figures can be interpreted quite differently. Much of the Indian higher-education system is in fact working for the industrialized countries. The massive outflow of scientists (40–50,000 to the United States, more than 10,000 to Britain and Canada) is a reflection of the country's extremes: an overdeveloped scientific system lost in a vast technologically underdeveloped infrastructure. Despite its atomic bomb and its missiles and satellites, the gulf between the enormous number of scientists and the high proportion of illiterates (almost 50 percent in a population of almost 1 billion) remains one of the most revealing contradictions of underdevelopment. However large the scientific community compared with all those of the Third Worlds, it seems a tiny enclave in the archaic world of the subcontinent, which is home to 600,000 villages speaking a language and leading a daily life totally untouched by the knowledge of that community. It is no coincidence that the Seventh Plan launched by Rajiv Gandhi in 1986 stressed that science had still made no real impact on rural development in the country, in spite of the growth of agricultural production.

Expanding the higher-education systems of developing countries runs the risk of exacerbating this distortion of supply and demand, by increasing the numbers with skills that their economies cannot absorb. From this angle, the kind of scientific education found throughout the Third Worlds, based on the model of the top institutions in the West, is obviously the least suitable for local conditions, in terms of its content as well as its aims. At best it allows 20 percent of an age group to receive higher education, but the other 80 percent leave school without an appropriate technical training—or else with none at all.[28] Science by its very nature is an elitist institution: Science in the poor countries is no more geared to serving the needs of the poor than it is in the rich ones.

Part 2
The Information Revolution

> We possess atomic energy and we also use cow dung.
> —*Jawarhalal Nehru*

6 The Looking-Glass Race

"It's a great, huge game of chess that's being played—all over the world—if this *is* the world at all you know. Oh, what fun it is! How I *wish* I was one of them! I wouldn't mind being a Pawn, if only I might join—though, of course, I should *like* to be a Queen best."[1] Alice describes very well the global game of science and technology being played at the moment, a game most developing countries find is happening "through the looking glass." It is a race in which the rich countries are always running faster and therefore further, whereas the faster the poor countries try to run, the more they find they are (at best) standing still.

Some of the Third Worlds have made great efforts since the 1960s to build up their scientific and technological resources, and the same is true of the international organizations that have helped them to draw up and implement sensible policies in this area. Yet the successes can be counted on the fingers of one hand, and they are in any case only relative in comparison with the enormous challenges of development.

As we shall see, the newly industrialized countries stand out as successes where the origins, the circumstances, and the possibilities for action seem exceptional, but their example does not offer a model that could be applied universally. On the contrary, it proves yet again that for the other Third Worlds, development is not a simple matter of a lag to be overcome, of an injection of science and technology, or of leapfrogging in the race to industrialize. These countries aside, the contribution of most developing countries to technical innovation—and thus even more so to scientific research—remains marginal, which is hardly surprising and, in any case and more distressing, has minimal impact on their own development.

GRADE, the Lima-based group for research into development run by Francisco Sagasti, from time to time publishes studies of R&D efforts in Latin America based on data and indicators gathered and processed with great critical judgment. In the most recent assessment, Sagasti notes that—in spite of the massive expansion of human resources over the last thirty years, the growth of investments in research, and the new opportunities provided by world trade in technology—progress seems infinitesimal, and the difficulties ahead appear gigantic. For all that it has 2.5 percent of the world's scientific and technical professionals, Latin America still accounts for only 1.8 percent of world R&D and 1.3 percent of the authors of scientific books and papers. It also has 8 percent of the total population and 5 percent of the gross output of the planet.

Again it is important to highlight the vast differences among countries. In 1982, Brazil alone accounted for almost one-third of Latin America's researchers (105,000 in all), and its research budget of $1.2 billion was double that of Argentina, three times that of Mexico, five times that of Venezuela, and six times that of Cuba. In the same year, the Andean countries together (Bolivia, Chile, Colombia, Ecuador, Peru, and Venezuela) had 24 percent of the region's scientific authors (1,754 out of a regional total of 7,450 and a world total of just over half a million). Since then, these disparities have not altered, except that the situation of the Andean countries has worsened.

"Difficult times," as Sagasti concludes. The efforts made in the 1980s were far greater than in the previous decade, but they yielded even poorer results. The same picture would be true of other developing regions besides Latin America. As the Queen says in *Alice,* you have to run as fast as your legs will go just to stay on the same spot, and run twice as fast to get anywhere else.[2] Through the looking glass, movements in time and space are so little governed by ordinary laws that the normal logic appears to be topsy-turvy. In the world of Lewis Carroll (a first-rate logician), life is unreal, like a dream, whereas dreams are more plausible than real life. But the world in which science and technology are now the factors that determine the military and economic might of nations is not a dream. To be more precise, it puts a stop to dreaming.

The Economics of the Intangible

Information technologies infiltrate and spread through social structures in extremely complex ways that defy any general description. By defining information as a basic resource needed for technico-economic activity, on a par with matter and energy, we highlight the role it plays in all aspects of the lives of individuals, groups, and societies. It is, however, still far from feasible to assess the different forms, means, and routes whereby the new technologies speed up, increase, transform, and renew not only information transactions but also the nature of production and consumption.

The kind of influence the information technologies have obviously depends on the context and the type of equipment used. The cost and therefore the efficiency of the equipment in turn depend on the level of know-how available as well as the use to which the chosen machines are put. In other words, it is doubly essential to avoid generalizations in this area: There is not just *one* Third World, there are several.

The new technologies penetrate, spread, and are absorbed in a great variety of ways, depending on the circumstances. Insofar as each country's situation is the product of a particular culture—India is not Brazil, Morocco is not Malaysia, and so on—each one creates its own history and culture under the influence of the new technologies. Nonetheless, despite the differences among countries or groups of countries, the vast majority of the Third Worlds are in the same discouraging position vis-à-vis the expansion of the new technical system. Because they are outside (if not totally foreign to) the intellectual foundations of this system, they are condemned to suffer its effects passively.

More serious still is that the modern technical system tends to undermine the importance of the one asset these countries had in their negotiations/confrontations with the industrialized countries: raw materials. The imperialist economic model of the nineteenth century was founded on acquiring huge reserves of raw materials and energy, and this model survived until the 1950s. The growing importance of information at the expense of energy and materials is reflected in production that is increasingly demanding in terms of intangible inputs and increasingly economical with energy and physical inputs. That being so, the South's exports of primary products to the North are less and less crucial to the prosperity of the industrialized countries.

The fall in oil prices was in fact not just the result of successful policies to find substitutes and to manage energy resources more efficiently: The West's response in this way to OPEC's efforts to fix oil prices was responsible for barely half the savings on oil. Much more important for the strategies to break the link between economic growth and oil consumption was the restructuring of the economy around and as a function of the new position of information. The whole industrial system had to adjust to a new situation in which consumption of nonrenewable resources was less significant.

When the U.S. economy came out of recession and started to grow strongly, the prices of many primary products should have risen. In fact, prices did not rise and some even fell, because of the restructuring of the economy. That is the explanation of Philippe Chalmin and Jean-Louis Gombeaud, in particular. They show that the amount of copper required to produce one unit of GDP in the OECD area fell by 20 percent between 1973 and 1984, nickel by 30 percent, and tin by 40 percent.[3] There was now a surplus of raw materials, and the cartels of producers were not in a position to influence prices.

Producers were weakened even further by the fact that the very definition of "raw material" was changing. For one thing, the materials that were increasingly important for the industrial system were becoming less and less "raw." The new materials were composites, produced from the commonest elements like silicon and carbon. For example, tungsten is no longer needed for the filaments of electric light bulbs, and the use of optic fiber made from glass will drastically reduce the demand for copper. The composites are made from a matrix that is then modified so as to give it the properties required in order to make each type of product. As this technology improves, it will be possible to make an ever wider range of new materials, and these will be increasingly made to order from ingredients that are available in unlimited quantities.

The example of Japan should have alerted people earlier to the possibilities of an economy built on intangibles: Isn't the most dynamic industrial nation also the one least well endowed with raw materials, oil, and space? Since 1973, in all the agencies of the United Nations system, the developing countries have fought for "a new international economic order" based on scarce raw materials, which was not simply a war that was out of date, but a whole culture. At the same time, Japan—without iron, coal, or oil—was building the most flourishing steel industries in the world and was also choosing to restructure its whole economy around information technologies.

Behind this decision lay the conviction that the economy of the future would be based on human rather than natural resources. In other words, because the new technical system ensured that information would be more important than matter and energy, it would make sense to adopt a strategy of innovation instead of relying on a policy of maintaining stocks. Technical education and training would doubtless play the same role in mastering the new technical system that physical resources had in the previous one—that is, the conquest of Lebensraum (living room) and raw materials. Japan was the first to understand that the transformations generated by the information technologies would lead to a restructuring of the entire economy around the production and consumption of completely new goods and services. Furthermore, mastery of all the aspects of the new technical system would be required in order to shift from a society concerned mainly with manufacturing activities to one in which service activities predominate.

Mastery of Use and Mastery of Production

In outline, two elements may be distinguished in the transformation of manufacturing and services by the information technologies: one concerned with the application of these technologies to various branches of economic activity, the other concerned with developing actual production of the equipment needed, whether "hardware" (the value derives from a physical object) or "software" (the value derives from intangible properties).

Two categories of technical mastery correspond to these two elements: mastery of *use* (i.e., the ability to integrate the information technologies into other activities, especially economic ones, so as to increase their productivity and competitiveness) and mastery of *production* (i.e., the capacity to adapt the productive system so as to produce information-based goods and services).

This distinction between the stakes involved in the mastery of use and of production is neither new nor peculiar to information technologies; it is related to the way the technical system develops and, in particular, to the international division of labor. It is inevitable that many more countries benefit from innovations than contribute directly to their design and launch. There are products some countries prefer to buy, either because they are not interested in producing them or because they lack the means to do so. In transportation, for example, mastery of use of air transport is quite widespread, whereas production of aircraft, especially long-haul carriers, is limited to a very few countries.

Industry based on innovation means having the necessary structures of research, development, management, marketing, and sales, and that in turn inevitably means concentration. The new technologies, especially those dealing with information and telecommunications, grow most rapidly in countries that have both a large well-qualified scientific labor force and an institutional infrastructure (fiscal, financial, and banking) disposed to take on very high risks.

Although at the outset innovations are more likely to start in small or medium-sized enterprises, the need to defend market positions usually leads small firms to ally themselves with larger ones. Because the life span of products created by the new technologies tends to get shorter and shorter, the industries that manufacture them must both maintain their innovative edge and try to market the products as widely as they can, if possible worldwide. Along Route 128 near Boston or in Silicon Valley, there are ultimately very few young firms that have managed to gain access to world markets without linking up with a larger firm or bank.

In theory, mastery of use of the information technologies is not only within the grasp of all countries but indispensable. The information technologies are everywhere penetrating and spreading throughout the economic and social system, affecting activities of all kinds, both in production of physical goods and in services. It would be impossible today to base the economic policy of a state or the commercial strategy and management of a firm on a rejection of innovation technologies. The only questions are to decide which technical solution should be chosen and how far to go in computerizing the firm, the sector, or the society.

Mastery of production of these technologies, on the other hand, cannot become as widespread. Under the previous technical system, based on energy and matter, only a few countries and firms were able to develop and run large plants or grids. The novel feature of the information technologies is that even to develop small-scale machines or programs requires such highly specialized labor and infrastructure that access to production remains strictly limited.

Even most industrialized countries possess only a fraction, sometimes quite a small one, of what is needed to produce the computer-related plant and facilities their economies need. This situation is even more apparent for the developing countries. In fact, only the newly industrialized countries are in any position to share in the production of information-based goods and services, whether of their own designs or merely copies. These disparities between the capacity to produce and to use can indeed be expected to widen, *even among the industrialized countries.*

The general tendency toward concentration of capacity in design and production promises to become particularly marked as a result of breakthroughs in the design of components and circuits. Throughout the whole process from R&D to production there are numerous possibilities of feedback, so that breakthroughs in manufacturing tend to influence the orientation and nature of research programs, while inventions and discoveries have an impact on how production develops. For instance, the LSI and VLSI (large- and very large-scale integration) circuits are so complex that they cannot be created without very large computers, and in turn the design of supercomputers requires the development of increasingly complex circuitry. The industrial firms that manage to master both aspects of this process therefore possess a substantial advantage over those that have mastered only one; this is one of the key factors in the lead that Japanese industry has gained in the area of the new generations of artificial-intelligence machines.

The same is true of program packages. They can be designed individually, which can be a rewarding task for small firms, or factories can be set up to produce them, using teams of specialists and libraries of reusable programs in conjunction with a suitably modified telecommunications system—productivity then grows by several orders of magnitude. Software production and sophisticated robotics can also be linked, so as to go even further in computerizing the manufacturing process. To master this aspect of the information technologies—and even more so the production of the new technical systems required for modern telecommunications networks—is clearly out of the question for many countries because of the high level of skills, organization, and financial cost involved.

While it is true that self-generating progress of this kind can be identified among the stages of technical change in the past, particularly since the Industrial Revolution, it has never been so intense or so systematic. The very nature of the new technologies thus contributes directly to making the international division of labor more differentiated than ever, as certain innovations are limited to a few countries and even to a few firms. The increasingly scientific character of the process of industrialization is tending to widen the gap between leaders and laggards: *Mastery of production is less and less related to mastery of use of the new technologies.* Just as there are fears that a "dual society" may develop in the industrialized countries—in which disparities in professional knowledge, position, and status will replace inequalities of birth—so there are risks that the world's "dual geography" will become more marked, with only a handful of countries (sometimes only a handful of firms) able to produce the high-technology products and services that determine comparative advantage in a sector of activity or even the future of the whole economy.

What possibilities will be left for the countries (even industrialized ones) that are eliminated from this race? In theory, what will matter will be the diffusion of technological innovations through the productive system and the ability to exploit them. After all, the whole history of technical change since the Industrial Revolution shows that no country ever manages to stay permanently in the lead. What is more, it suggests that it is often more economical to come along after the inventors, and buy their inventions and learn to use them, than to take the risks involved in being one of the leaders. The countries in the strongest position should therefore be those that are the most skilled in taking advantage of new systems of machines, not those that design and produce them.

The example of Japan up until the 1970s ought to be comforting for those countries that prefer to buy technologies from others and settle for being imitators rather than inventors. Since then, however, Japan has chosen quite a different strategy, having discovered painfully that even if diffusing the innovation is what counts, access to it is necessary first. And technical change is now passing via unavoidable points—custom chips and programs—where the inventors can retain their monopoly. These custom chips are critical for the latest computers, and as improvements are made in integration, an ever greater number of elements of a circuit can be integrated into one chip, and the component itself becomes a system.

The major producers of computer and telecommunications systems (particularly AT&T and IBM) used to manufacture for themselves most of the custom components they need. Firms or countries that do not manufacture these components but want to use them find they are unable to acquire the necessary expertise and therefore cannot even use them. A custom chip contains concepts and data that determine the productivity of the final product, so that robots designed by computers using these chips enjoy an unshakable and decisive comparative advantage, not just in computing power but above all as regards automation of production.

It is not by chance that the Japanese are investing so heavily in encouraging local production of complex components; they are hoping to break the U.S. monopoly in this area. The United States argues that the dual nature of these technologies—although they are meant for commercial use, they may also have military applications—means that access to them should be restricted, even for U.S. allies. The strategic significance of these technologies is not merely military, however: From an economic standpoint, firms or countries able to retain control over production have an unassailable lead in the mastery of use of the new systems.

Any country that depends on another for its competitive position is by definition vulnerable, but when the capability to apply the innovation is directly linked to the capacity to design new products, the room for maneuver is even smaller. Access to the components that are necessary in order to design the products thus becomes the sine qua non for diffusing the innovation and mastering its applications.

The Computing Capacity of the Third Worlds

The Information Revolution has above all made the poor countries more dependent than ever on the rich ones. The new information technologies in the narrowest sense (those that process information) are far from scarce in some developing countries, but in the vast majority of cases it is quite beyond their capacities to engage in the production or (even more) the design, which happens in the labs and private firms of the industrialized countries.

This is even more true of information technologies in the wider sense: supercomputers, telecommunications networks, satellites and other space technologies, conventional weapons using advanced guidance and control systems. A whole section of the new technical system separates the industrialized countries from the developing world as radically as the guns and steam engines of sixteenth- or nineteenth-century Europe set it apart from the continents then under its domination. Mastery of production of the new systems puts the countries of the North in such a strong position that they need not worry unduly about the recent progress made by the newly industrialized countries. As for mastery of use of the new technologies, that is distributed in line with the different societies' capacity to receive and absorb them.

The computing capacity of Latin America was built up in the 1960s, at the same time as that of the industrialized countries. The absence of local skills and the complexity of the new industrial sector explain why at first all the necessary equipment had to be imported. In fact it was not until the new electronic systems had been standardized and miniaturized in the 1970s that certain countries were able to consider creating their own computer industries.

Aside from the subsidiaries of the multinational corporations, the main user of computers from the moment of their introduction was the state, for its administrative services and for industries in public ownership. In most cases, these computers were installed in major data-processing centers, so that virtually all the equipment and trained personnel were concentrated in one place. This tendency clearly arose from the nature of the huge main-frame machines of the 1960s. Initially, the developing world's stock of big, all-purpose main-frame computers was almost entirely North American (90 percent), with IBM alone accounting for 63 percent.[4] The remainder were either British (ICL) or French (CII), as these countries exported to the captive markets of their former colonies. In the first decade of computer expansion, by far the best-equipped region of the developing world was Latin America (58 percent); Asia accounted for 28 percent, the Middle East 8 percent, and Africa just over 5 percent. Moreover, a very few countries possessed the bulk of this stock: In Latin America, the four leading countries in descending order of importance were Brazil, Mexico, Venezuela, and Argentina; in the Middle East, Iran, Egypt, and Turkey together owned half the region's stock; in Africa, Algeria, Namibia, Nigeria, and Zambia made up 54 percent of the total. The annual growth rate has been very rapid over the last two decades, but the overall distribution has not changed significantly.

Because information technology in developing countries has been adopted mostly in public administration and in large firms (public and private) with access to international markets, the applications have therefore been confined largely to the needs of management and basic administration (payrolls, accounting, taxation). By automating administrative procedures, state agencies and major firms have been able to acquire the practices and standards of the world economy, but—unlike the experience of the industrialized countries—this type of computer use has not generated any new impetus in other sectors. In many countries, in fact, the hiatus between the collection and analysis of economic data has meant that these machines have not been used for economic forecasting, planning, and monitoring. Moreover, data banks in developing countries have mainly been assembled by international organizations (e.g., FAO, WHO) or by private foreign companies.

Telecommunications undoubtedly provides the most efficient way of decentralizing computing capacity. However, the installation costs of national networks of computer-based telecommunications and the poor quality of existing telephone systems in many developing countries are responsible for the difficulties that have been encountered in efforts to disperse major computing capacity. Only a few institutions—public or private—have been able to set up their own networks (airlines, banks, the police).

Telephones are indispensable (at least until deficiencies in the land networks are overcome by satellite) because it is via telephone links that it is possible to have access to data transfers, interactive communications, videotext, electronic mail, and databases, both nationally and internationally. In 1981, the industrialized countries had an average of 46 telephones per 100 inhabitants (in the United States, 92 percent of households have a telephone), whereas the average in the developing countries was less than 3 percent—Africa had 0.8 per 100 inhabitants, Asia (excluding Israel and Japan) 2, and Latin America 5.5. In addition, a high proportion of telephones in these countries are concentrated in large urban areas: In Mexico, for instance, there are still more than 30,000 communities without any access at all to a telephone, and the situation in the rural areas of China and India is very similar.[5]

Access to television is just as uneven—90 percent of the receivers are concentrated in the industrialized countries (with 15 percent of the world's population), leaving just 10 percent for the developing countries. In 1980, one person in 500 had a television set in the developing world, as against one in two in the industrialized countries.[6] A television screen gives access not simply to the broadcasts of television programs (via conventional means, cable, or satellite) but also to the use of computers, videos, and, in conjunction with the telephone system, a whole range of data flows. However much declining costs and technological progress help to facilitate access to the new technical system, the spread of information technologies in the developing countries must first overcome the barriers of low income and poor infrastructure.

In the 1970s, a handful of developing countries began to establish policies on information technology aimed at reducing their dependence on foreign expertise by developing their own production capacity. These policies dealt simultaneously with the technical, academic, industrial, and regulatory aspects (including the control of imports and support for local enterprise), the objective being the creation of a full range of equipment and systems. Very few developing countries have the financial resources, the institutional infrastructure, or (above all) the qualified personnel to institute policies of this kind. In addition, the technical experts would need the support of politicians who are aware of the potential benefits of information technologies and who are both nationalist and interventionist; experience of the major technological initiatives suggests that this factor is far from being the least important.

Among the first of the countries to adopt this approach were Brazil and India. In 1984, the Brazilian computer industry, with a turnover of $1.5 billion, accounted for less than 1 percent of the world total, but its output nevertheless represented 75 percent of the total for Latin America and more than 59 percent of Third World production. The Brazilian effort dates to 1972, when a committee was formed to coordinate the design and construction of the first Brazilian-made computer by the two universities of São Paulo (for the hardware) and the Catholic University of Rio de

Janeiro (for the software).

Two years later, the government decided to launch a national computer industry, and the firm of Digibras was set up with two-thirds of the stock in Brazilian hands (one-third public and one-third private) and the remaining third held by a foreign company, Ferranti. Digibras in turn created the firm of COBRA (Computadores Brasilerias), which supplied medium and small computers for government and industry. Digibras itself evolved into an agency for evaluation and encouragement of R&D and became a holding company involved in the manufacture of components, the backup for the national network for data transfer, and the regulatory body controlling contracts for the acquisition of equipment by the federal government.

The policy was confirmed and codified by a law (October 3, 1984) defining which markets for mini- and microcomputers were to be reserved for Brazilian firms (those whose capital is held entirely by Brazilian interests). This meant that not only were multinationals excluded but so too were joint ventures between Brazilian and foreign firms, even if the latter owned only a minority share. The United States, under pressure from IBM, protested constantly about this deliberate infringement of the rules of free trade, but the Brazilians justified the law by referring to the Japanese and even U.S. examples (the "Buy American" Act), because articles 18, 20, and 21 of the GATT agreements authorize such measures in two instances: for reasons of national defense and to protect infant industries.

The result is that the national element in the Brazilian computer industry has been considerably strengthened relative to the predominance of foreign suppliers of big machines, especially U.S. (above all IBM) but also Japanese and French. Between 1980 and 1984, the share of locally produced mini- and microcomputers in the national market grew from 17 percent to 95 percent. The price differential, for the same quality of machine, between Brazilian products and those of the industrialized countries has been diminishing steadily, though it remains large and (because the quality is not in fact always the same) there is a substantial black market in smuggled imported products. Nonetheless, Brazil has become a major exporter of these smaller machines (earning $250 million per year from them), and even though the net balance of trade in computers is still negative, the exports do contribute to foreign exchange earnings.

Brazil ranks third among countries that supply their own computer needs, after the United States (90 percent) and Japan (54 percent). The local manufacturing side employs on average twice as many staff as the subsidiaries of the multinationals and seventeen times as many in R&D activities. The 1984 law thus acted as a stimulant to modernizing local industry and to developing skills needed for the technical support and the expansion of the universities, and it ensured a more "equitable" share in international trade by favoring the export of manufactured goods rather than primary products.

The Brazilian approach was not unlike the French policy of twenty years earlier and in any case was inspired by the same strategic concerns. Just as the French "Plan Calcul" was the result of a desire to protect the independence of the new industry for reasons of defense (the U.S. government had refused to supply a powerful computer for the French strategic forces), the Brazilian initiative was guided by engineers trained at the military academies who were concerned simultaneously to modernize the army and to expand national industry.[7] Technological nationalism and interventionism tend to go hand-in-hand: Mastery of production appears as a guarantee not only of political independence but also of access to technologies in the future. This perspective explains why Brazil's president described SERPRO—the federal agency responsible for data processing for the whole country, simultaneously the tax-collecting body, the foremost customer of the national computer industry, and one of the main centers for training in computer skills—as "the laboratory of the future."[8]

Unlike France, Brazil has not attacked the territory of the major U.S. manufacturers head-on. Instead, it chose the profitable market opportunities neglected by the big corporations: mini- and microcomputers, terminals, peripherals. The market in these items is strictly reserved for the Brazilian firms that grew out of the first links with foreign companies. The market for big (and very big) machines is open, though imports are still tightly controlled (IBM 56 percent; Burroughs 15 percent; Bull in association with the Brazilian firm ABC 3 percent).

Since the end of military rule, Argentina has been trying to conduct a similar policy. In 1985, "Resolution 44" set out a plan intended to create a national computer industry based on joint ventures that would receive preferential tariff status. Anxious lest others copy this and the Brazilian initiative, the U.S. government exerted strong pressure quite openly to make Brazil abolish the law on reserved markets, which threatened IBM's policy (no technology transfer to third parties but instead local production by local subsidiaries) and interests directly. A struggle ensued in the shadow of negotiations about the loans the International Monetary Fund (IMF), with U.S. backing, was prepared to make to Brazil. The reserved market has been abandoned.

This interventionist policy had undoubtedly positive results for the domestic economy and for Brazil's share of international trade. The Brazilians were all the more enthusiastic about it because their success with computers illustrated the success of the efforts to speed up industrialization in general. Nevertheless, the policy also had its limits, as the strict control of imports undermined the expansion of certain sectors of the economy as well as some defense technologies, which could not make use of the most advanced systems available. The 1984 law was toned down, above all with regard to software and telecommunications, so as to allow jointly owned Brazilian and foreign firms to be created once more and encourage local industry to catch up.

Very few developing countries could hope to adopt a policy like that of Brazil or India aimed at creating and protecting a relatively independent national industry. Lacking the size (and hence an adequate domestic market), the financial means, the university resources, and the trained labor force, as well as the strong motivation that underpins such an interventionist strategy, mastery of production is beyond the grasp of most developing countries. At best, they can hope to gain a small share in the know-how via the subsidiaries of multinationals, whose fate is determined abroad. Most are condemned either to import directly the "black box" of up-to-date technologies, designed and manufactured in the industrialized countries, or else to assemble or sell products from different components designed elsewhere.

Even the big countries, for that matter, may fall into the black-box trap, as is clear from the report of a former director of a research center in Beijing: From 1984 onward, China imported enormous quantities of video-display units and components, at a cost of $300 million, but most of the 100,000 computers for which these parts were intended turned out to be unusable because of lack of supplies, peripherals, programs, programmers, and operators. It would have taken at least five years to train the 100,000 technicians required to make use of the machines, and by then the computers would have been obsolete. The author concludes that one reason for this failure was the poor coordination, information, and consultation of the state agencies at the national level and the inadequate management by local authorities.[9]

The extraordinary expansion of microprocessors and the development of space technologies have doubtless modified the situation for many developing countries, making policies for exploiting computers and telecommunications more viable, at lower cost, and more suited to local conditions. The smaller countries (South Korea, Taiwan, Hong Kong, Singapore, Malaysia) have specialized in producing components (circuits, terminals) for the multinationals to assemble into computers or systems. The larger or more securely industrialized countries (Brazil, China, India, but also Indonesia and Mexico) have extended their efforts into telecommunications by developing and launching their own satellites.

Aside from the three giants (Brazil, China, and India)—the only ones able to mobilize their resources across the whole range of the technical system, from computers to telecommunications, from rockets to satellites—very few developing countries have the possibility of independent production of even *some* elements of the system. Thanks to lower costs and the diffusion of more flexible systems, more seem likely to be able to benefit from computing and telecommunications. But the vast majority of the countries of the Third Worlds are, and will long remain, closely dependent on the industrialized countries, even for the application of new technologies to the solution of their own problems.

7 The Machines from the North

> My Lords: During the short time I recently passed in Nottinghamshire not twelve hours elapsed without some fresh act of violence; . . . I was informed that forty Frames had been broken the preceding evening. These machines . . . superseded the necessity of employing a number of workmen, who were left in consequence to starve. By the adoption of one species of Frame in particular, one man performed the work of many, and the superfluous workers were thrown out of employment. . . . The rejected workmen in the blindness of their ignorance, instead of rejoicing at these improvements in art so beneficial to mankind, conceived themselves to be sacrificed to improvements in mechanism.

In 1982, Wassily Leontief, Nobel laureate in economics, began an article for *Scientific American* on the consequences of the Information Revolution for employment with this quotation from Lord Byron. The article made a great impact, because Leontief's conclusions (based on computer calculations of the input-output model of the economy he had invented) were extremely pessimistic. When applied to Austria, his calculations predicted a higher level of unemployment in 1990 than that country had experienced in the dark days of 1930—a level that would remain high even if the working week were reduced.[1]

The debate about the threats of industrialization for employment is an old one, and it is a sort of ritual for all historians of the Industrial Revolution and economists concerned with technical change to allude to the past in order to deal with the uncertainties of the present: After all, the experience is, on balance, reassuring, isn't it? Leontief was merely participating in the ritual when he quoted Lord Byron. When Byron himself made his maiden speech in the House of Lords in February 1812, he was trying to explain and excuse the revival of unrest among workers who had lost their jobs when new machines were introduced. (His plea was not for nothing, because the House of Lords was debating the introduction of the death penalty for workers who destroyed machinery.) A generation earlier, Ned Ludd had incited his fellow workers to smash the knitting machines that were beginning to be installed in workshops; the Luddite revolt ended in bloodshed, like that of the silk workers in Lyon in 1831.

The Specter Returns

Many of the questions raised during the House of Lords debate in 1812 are being discussed again today. In the competition between human beings and machines, won't the machines surely win, carrying out their tasks more regularly and imposing their rhythm? At the same time, won't they lower wages, eliminate work, create unemployment where there was none before, and increase existing joblessness? The history of mechanization is marked by fears and curses that echo down to our own day, with barely any change, in connection with the information technologies.

These issues are far from new; they date even to before the Industrial Revolution, as is clear from Marx, who refers to a machine for weaving ribbons and braid (the *Bandmühle*) that could produce between four and six items simultaneously. This loom was developed in Danzig in 1529, but is said to have been suppressed immediately and its inventor "suffocated or drowned," the city magistrate fearing that "the invention would turn many workers into beggars." A century later—and thus well before the first machine tools and the steam engine—the same textile machine was used at Leiden and Hamburg, causing a riot of the haberdashers so violent that the authorities eventually outlawed the machine and had it burned in public.[2]

The first computers sometimes suffered a similar fate, with their introduction in some firms meeting the same sort of rejection. Despite the success of the Industrial Revolution, and although even the working class and the trade unions have learned to see machines as allies rather than enemies, the specter of workers being supplanted and made redundant has never entirely vanished. It has indeed reappeared in the last few years, all the more vividly because the rapid growth and full employment of the postwar years came to an abrupt end.

Across a period of a century and a half, hasn't history demonstrated the errors of the critics of technical change—Luddite workers and "romantic" intellectuals—who predicted pessimistically that jobs would be destroyed, leading to growing poverty and social unrest? In fact we have seen the opposite: The successive waves of technical innovation that marked the various phases of the Industrial Revolution from the steam engine onward have generated a spectacular increase in employment. However great the downswings of the cycles, the notions of relative and absolute pauperization have been abandoned to tattered ideologies.

Nevertheless, after the years of unprecedented postwar growth, the industrialized countries had to face economic problems that were expressed in, among other things, a general rise in unemployment that has proved lasting. It is difficult to identify a single culprit: inflation, lack of investment, the rise in oil prices, failures of macroeconomic management, and so on. It may well be, as Zvi Grilliches humorously suggested, that as was the case in Agatha Christie's resolution of the murder on the Orient Express, all the suspects will turn out to be guilty.[3]

Every economist has his or her own ideas about the diagnosis and the cures, but they all agree on at least one thing: Technical change—the Information Revolution—was neither the trigger nor the decisive factor in the development of the crisis. Many studies have been carried out in the industrialized countries and by the OECD secretariat to measure the impact of the new technologies on the level of employment. All indicate substantial job displacement, but none has managed to demonstrate a negative correlation between the spread of new technologies and the demand for labor. The most recent studies conclude that in current circumstances, aggregate employment depends on a wide range of factors unrelated to technical change. Curiously, however, while none of these studies has been able to show that the new technologies have caused a net loss of jobs in the OECD area, none has attributed to them any substantial *growth* of employment.[4]

Why, then, these fears, which began to be apparent at the end of the 1970s and which first-rate economists have echoed ever since?[5] The problem is that the past is no longer necessarily a reliable guide. Whereas formerly most major technical innovations replaced human strength in producing goods, now (and the trend will be greater in the future) they tend to replace human intelligence in both manufacturing and services. As Leontief points out, the relationship between the human being and the machine has been radically altered. Experience, as well as economic theory, tells us that in the long term, whatever the social cost of layoffs, job losses, and retraining, technical change generates an increase in wealth and new jobs. But will the lessons from an outdated technical system always apply to a new system?

For most economists, the impact on employment of computers and information and telecommunications technologies is not very different from the consequences of all the earlier innovations for the history of mechanization. For others, however (and not the least able—they include people such as Christopher Freeman and Wassily Leontief), there is a risk that the new technical system will entail a drastic reduction in labor needs. We are passing through major structural changes, and this may imply the exhaustion of the postwar growth model. Once reconstruction was over, Europe and Japan absorbed the patterns of production and consumption prevailing in the United States. All the indications suggest that the management model and the organizational structures brought about by the technical system based on matter and energy are no longer adequate.

The Information Revolution will increasingly influence these changes. For Leontief, it undermines the dominant position of human labor as the main factor of production; for Freeman, the new paradigm shows a bias in favor of saving labor rather than saving capital.[6] Both agree we are just starting this process: Unless appropriate measures are taken—political, social, and above all educational—the information machines will be increasingly economical in their use of labor. Neither is so pessimistic as to declare that these measures will not be found; on the contrary, they believe there are solutions for the industrialized countries. But what about the developing countries?

Man Cannot Live by Signs Alone

It is not enough to look at the example of the industrialized countries to foresee what will happen elsewhere. The way the affluent nations have already adapted to the "informatization" of society can indeed help to shed light on some of these problems, especially the nature of the obstacles and the resistance that the spread of the new technical system may encounter. But the situations are so dissimilar that the same causes cannot have the same effects or, above all, the same solutions.

The information technologies satisfy needs that were generated by the very way the industrialized countries have developed. It is far from clear that these technologies are geared to meeting the priority needs of most developing countries efficiently and rapidly. Yet at the same time, it is inconceivable that any of these countries should choose to do entirely without the products and the infrastructure that increasingly define the "nervous system" of the modern world and on which it depends in order to function. All in all, the rich countries have both the means and the time to adjust to the impact of change: However high their unemployment, they do not have to cope with high population growth as well. The poor countries have neither the time nor the means.

The information technologies were created and flourished in societies having features that are the complete antithesis of those in the Third World: substantial technological potential linked with highly varied industrial and academic structures; a relative shortage of labor; vast capital resources; a service sector providing more jobs than manufacturing and far more than agriculture. In addition, the traditional consumption patterns associated with consumer durables, based on matter and energy, have changed or reached saturation point; the demand for services has expanded as a function of both changes in these patterns and the growing complexity of economic and social organization.

In the United States, almost 70 percent of the employed population works in services, the same proportion as worked in agriculture a century ago. The same is true in Europe. After World War II, more than a third of the workforce was employed in farming, whereas this figure has now fallen below 10 percent in most countries.

These changes led several authors (such as Daniel Bell, Peter Drucker, and Yoneji Masuda) at the beginning of the 1970s to talk about the "postindustrial age" starting in the industrialized countries. The main features identified were a dominant service sector; an increasing intellectual element in economic activities; the shift from a centralized and hierarchical Taylorist model of work organization to one more flexible and decentralized; decline if not disappearance of the world of the working class in the nineteenth-century sense; the rise within the middle classes of new elites who create and use knowledge. In 1976, Marc Uri Porat wrote *The Information Economy,* a first attempt to measure the structural changes brought about by the growth of the information sector. Under his definitions, almost half the U.S. workforce in 1974 was working in information-related activities.[7] Since then, these changes—in which the United States yet again led the world—have spread to most of the industrialized countries.

The term "postindustrial society" does not explain everything, because even if the industrialized countries go beyond the nineteenth-century model of production and work organization, they are not really turning their backs on all that made them what they are. There is no sharp break between the process of industrialization and the Information Revolution; rather, industrialization simply uses the information technologies to extend further, through new forms, channels, and repercussions. Whether the work involved is physical or intellectual, it is still a matter of mechanization. Moreover, the measurement of information-based activity is open to criticism because it tends to combine data in such a way as to reduce all activities to their information aspects, even when they are still concerned mainly with manipulating goods or materials.

Whatever the reservations in this regard, it is an undeniable fact that the proportion of information-based jobs has grown substantially in the industrialized countries, partly because of the expansion of employment in the service sector, partly because of the replacement of blue-collar workers by a workforce more and more engaged in tasks linked in some way with information. The industrial system increasingly deals with intangibles: Even when it is producing things, human effort is essentially concerned with handling signs. In all sectors, the proportion of white-collar workers is rising steadily; today one in two U.S. workers—as against one in eight in 1900 and one in three in 1950—produces, processes, or transmits information. The proportions are roughly the same if one takes instead the share of value added by the information and communications sectors in total GNP.

The situation in the Third World countries is the converse. One of the characteristics of underdevelopment is indeed the absence of the features that underpinned the information-based thrust in the industrialized countries. Agriculture is still and will long remain the dominant sector; the population is mainly rural; education at best is limited to primary schooling; industry and universities are by definition underdeveloped; some cities are vast and sprawling, without the means to provide an adequate infrastructure; the service sector is unlikely to expand rapidly; there is insufficient capital; and the huge labor force suffers from endemic underemployment. With basic needs far from satisfied, and under pressure from both population growth and the worsening debt problem, the working conditions—when there is work at all—are not directly affected by the technical revolution occurring in the North.

Human beings cannot feed themselves by handling signs and symbols; an economy based on intangibles is the antithesis of what the South needs. All the questions discussed in the industrialized countries relating to the spread of the new technologies have a very different ring in the developing countries. And it is misleading—indeed, downright damaging—to suggest that these countries can hope to derive from the Information Revolution benefits similar to those that will help resolve the problems of the industrialized countries. The Information Revolution may allow some of the Third World nations to adopt strategies in order to catch up, but it does not in any way mean that any of them can bypass the issues of development—unless it is supposed that the main purpose of the spread of the Bible in Europe as a result of the invention of printing was to feed bodies rather than minds.

The Medium Is Not the Message

No technical system has consequences that are unavoidable. The effects depend on how societies themselves evolve, on how the new technologies are introduced, on how well the ground has been prepared for them, and on the relative strengths of the various actors involved, just as much as on the purely technical factors. Technical change is never cut and dried, and it is no more plausible to expect technologies alone, whatever they are, to overturn the power structures governing the relations among states than to cause radical shifts in the social positions of groups and individuals within nations.

More important, technical developments may themselves provide new responses to the problems they raise. From this angle, it is undeniable that some of the advances in information technologies (computers and telecommunications) seem to offer technical solutions that *are* well suited to the needs of the Third World. These advances do not merely involve the development of very sophisticated systems but also make possible improvements to conventional systems, reducing costs and increasing their efficiency.

A whole range of advances gives credence to the view that there are new fields to be exploited in which success does not depend exclusively on the technical and economic circumstances enjoyed by the most industrialized countries. These include increasingly user-friendly access to computers, operating systems independent of conventional electricity supply, and telecommunications via satellites that overcome the deficiencies of the land-based telephone network. Improvements to databanks and to information processing are making access to information technologies much easier in small, isolated communities with little in the way of infrastructure. Decentralized systems may encourage more efficient management methods in small and medium-sized firms, especially rural cooperatives, and provide new opportunities in farming, health, and education. The new and more flexible systems of data collection and processing are already bringing benefits to natural-resources assessment (teledetection), weather forecasting, and the avoidance of natural disasters such as the migration of insect pests. The development of more accessible expert systems promises to facilitate the distribution of medicines, help in the diagnosis of disease, and promote health education, the organization of health care, and epidemiological monitoring.

Yet technical advance is one thing, the social and economic setting in which it occurs is quite another. It is easy to give in to the temptation of cheap "futurology" and its emphasis on the *supply* of new technologies and technical systems that are more flexible, decentralized, and beneficial, as if they will automatically satisfy all the needs of the developing countries. This is to overlook the constraints that limit the potential *demand* in these countries, where some of the products that have reached saturation point in the rich countries remain indispensable for survival, and where some of the new products being imported are particularly ill-matched to consumers' needs because they require suitable, advance introduction into the culture.

As regards education, common sense would remind us that computers use languages and codes that presuppose at least the ability to read and write. And it is foolish to talk about computer-aided education (or other forms of assistance) as a panacea; the experience of the industrialized countries suggests a more ambiguous, if not downright disappointing, assessment. The medium can undoubtedly become the message, but the machine's abilities are never a substitute for the effort of acquiring knowledge. Machines can assist the learning process, and teaching programs are useful pedagogical tools, but neither can replace the teachers.

Everywhere, the bottleneck to "informatization" arises not from the supply of technologies but from the scale and the nature of the demand, which depends partly on the level of average incomes and partly on the availability of a labor force trained to use the new technologies. Access to the new literacy does not mean any savings with regard to the old. Computers are far from being a substitute for education and instead require a type of education geared to spreading familiarity with the new technologies—hence the efforts being made by many industrialized countries to raise themselves to the level of the "new culture" and the high priority many give to education and training policies that will meet the challenges of the new illiteracy. This means acquiring the ability to handle information and communications systems in every sector, not just in connection with producing certain machines and software but also in using the algorithms and grammars on which they are based.

For most developing countries, the projects to introduce the mass of the population to these skills and to offer appropriate qualifications are bound to be utopian, partly because circumstances are not propitious and partly because they lack the means. To "informatize society" means first making computers more sociable, which would require massive investment in machines and personnel.

Between Nightmares and Utopian Dreams

Amid the uncertainties of the final years of the twentieth century, the new technologies offer both hopes and fears of a new age: They are credited by turns with bringing about a radical transformation of society, starting a new era in the relations among groups and individuals, ensuring a dominant place for imagination and conviviality, or else of providing the means to carry out technocratic and totalitarian plans—of being the indestructible tool of either traditional tyrannies or new imperialisms and marking the ultimate in the division of the world into haves and have-nots.

The prophets in the industrialized countries range in perspective from the deepest pessimism to utterly utopian optimism. The pessimists stress the risks of centralization, bureaucratization, and trivialization of activities, not to mention the threats to privacy, with the shadow of Big Brother watching over a universe of robots and increasingly elitist and totalitarian social organization. The optimists dwell on the miracles around the corner, on the growing independence and decentralization that will enable the "weak" and the "little people" to withstand the economic and political Leviathans and then to flourish in a creative environment with every opportunity for leisure and cultural activities.

The main issue, however, is employment. Here the controversy between optimists and pessimists generates the wildest bets. The pessimists concentrate on the short-term losses, whereas the optimists look to gains in the long term. In the absence of a reliable model and a satisfactory theory of technical change, the only way to analyze the real and the likely effects of information technologies on employment is by resorting to drastic simplifications. And as the effects are not evenly distributed in time and space, there is no assurance that the current experiences of the industrialized countries will be repeated in the developing countries.

These reservations are all the more necessary because the degree of responsiveness to these effects varies so widely, even among the industrialized countries. The United States and Japan have tended to view the alarm raised by the links between information technologies and unemployment as largely psychological in origin; Europeans have been more hesitant, worrying not just about the overall impact on employment but also about the disappearance of certain skills, the effect on qualifications, and the relocation of activities.

In this revival of the old debate about the links between mechanization and employment—a debate as old as economics—the difficulties that followed the 1974 oil crisis (we do not linger on whether the one caused the other) led to questions about the role the revolution in microelectronics might have played in the massive increase in jobseekers during the 1970s, from 10 million in the OECD area in 1970 to over 30 million in 1986 and into the 1990s. Nothing conclusive emerged, but the contradictory interpretations proliferated, and even greater uncertainty surrounds what is likely to happen in the future.

The economic crisis appears first of all to be a crisis of economic thinking, in that the Keynesian notions that accompanied the unprecedented growth rates after World War II proved inadequate in the face of the combination of rising unemployment, runaway inflation, and very moderate rates of growth. Some commentators even suggested that the war itself had been the most efficient remedy for the unemployment of the 1930s. The ineffectiveness of the remedies currently proposed, from fine tuning to stimulating demand or monetarism, confirms that the crisis cannot be dismissed as just a short-term downswing and that instead we must deal with structural problems in the long term. The gigantic U.S. budget deficit and the fluctuations in the dollar exchange rate do nothing to help restore world economic equilibrium.

In the structural transformations that are taking place, it is impossible to measure the part that can be attributed to the recent spread of information technologies. Even if none of the many studies of the subject can show that technical change was in the least responsible for triggering the crisis, it cannot be ruled out that it will exacerbate the results in the medium term.[8] As the economies of the industrialized countries shift increasingly toward the service sector, it is clear that agriculture and manufacturing offer fewer and fewer jobs in order to produce an ever greater quantity of output. Hence arises the question that nobody nowadays can answer without appearing to consult a crystal ball: If the service sector is going to continue growing, can it generate enough jobs to offset those lost in agriculture and manufacturing, given that it is itself likely to see considerable productivity increases thanks to the information technologies?

Among the other reasons mentioned, probably rightly, for the fall in employment is the rigidity of social structures, and indeed resistance to the destabilizing effects of technical change does tend to increase during periods of structural adjustment. Jobs are lost in those firms, sectors, and regions that cling to the social structures of the past, and they grow wherever there is innovation. The "deindustrialization" of the United States can be seen in the shift of activity from the East Coast to the West, from sectors based on processing matter and energy to the new ones linked to information. Europe is still shedding jobs in the industries inherited from the early phases of the Industrial Revolution so as to stimulate those that promise to determine success in the future. In this process of modernization, Japan has forged ahead by choosing early on to sacrifice the lame sectors and invest in the new technologies.

The conflict between the sectors of the past (mining, metalworking, shipbuilding, textiles) and those of the future (electronics and high tech) illustrates the shift into the latest phase of industrial capitalism, forcing economists to look again at the issues that were dear to Schumpeter. According to Schumpeter, the process of creative destruction is the most characteristic feature of industrial capitalism—which is driven forward by new products, production methods, markets, forms of organization, and consumer tastes. These great waves of innovation lead to major breaks, followed by recessions, adjustment, and upswing in cycles of varying lengths. As they try to estimate how long the recession will last, economists read (or reread) Schumpeter to find out how technical change influenced the fluctuations in the course of the Industrial Revolution. They disagree about the existence of long cycles, but one does not have to believe every word of Schumpeter's theories to see the plausibility of the link between periods of major innovation and the difficulties of adjustment in various sectors of the economy.[9]

Experience and Theory

The spread of information technologies through the economic system—whether to produce material wealth or to provide services—generates two types of macroeconomic effects. Existing sectors are transformed as machines replace human beings, which leads to job losses in areas where human intelligence is used for repetitive tasks, either combined with physical strength (e.g., the assembly-line worker) or for purely information-based purposes (e.g., manual bookkeeping or typesetting). In addition, new jobs and productive activities are created, which affect the design, construction, maintenance, and operation of machines or languages for handling information.

Both the experience of the industrialized countries and economic theory—in fact a theory that goes back to David Ricardo—show that in the long run the new jobs created will offset those lost as a result of technical change. Improvements via mechanization will bring down prices, so that all consumers (even the least well-off) will benefit. In theory, even if the productive capacity of a sector increases faster than demand, the workers made redundant will be absorbed in other jobs.

Because technical change alters both the pattern of demand for skills and the industrial structure, the retraining of the labor force must keep pace with the spread of the new technologies. Herein lies the risk of major structural unemployment, as the job opportunities are reduced for unskilled younger workers in general, for older and less mobile workers in traditional production and sales, and for women without higher education or qualifications in search of new jobs. In these categories, job cuts may continue for a long time, while new jobs are created in all the new sectors, for designers, programmers, engineers, maintenance specialists, and the like.

Since the beginning of the Industrial Revolution, the increase in productivity arising from mechanization has been accompanied by an increase in the active population in all the industrialized countries. It is necessary to assume that output remains constant despite technical change for the increased production per worker to lead to a fall in the numbers of workers required. In fact, improvements in productivity cause the costs of production in real terms to fall and incomes to rise, leading to growth of both demand and output. Consequently, the application of new technologies should make it possible in future, just as in the past, to draw on new resources that will in turn increase employment.

This brings us back to the question, How long will the transition period last? It is impossible to give an answer now, as there is no way of knowing how long it will take for the new jobs created to offset those lost. At the level of the economy as a whole, the compensatory effects are linked to the lag before the new technologies stimulate new investment and, as an indirect consequence, disposable income and demand increase. They are also related to the way in which labor and education policies manage to take account of the foreseeable effects of technical change. In fact, this process is rather more complex, as each firm, sector, or country responds differently, depending as much on its capacity to cope with competition as with the direct impact of technical change. The internationalization of production and trade explains why, because of these differences in competitiveness, the compensatory mechanisms operate very unevenly.

Nevertheless, the moral to be drawn from experience as well as from theory is reassuring: In the short term, the transition period creates major difficulties for those who lose their jobs or fail to enter employment because of redundancies or lack of skills; in the long term, the number of jobs created should offset those lost. In the meantime, it is impossible to turn away from the new technologies. Preventing the use of more productive machines and methods, when foreign competitors do not, would probably have even more disastrous effects on domestic employment. The industrialized countries have no choice but to "run ahead" of technological progress. To seek to conserve employment by limiting the introduction of the new technologies would be counterproductive. In the current climate of sluggish economic growth everywhere and of increasing international competition, the only way to overcome the difficulties of adjustment and the social unrest they might cause is to step up efforts to ensure that the labor force has adequate opportunities for training and retraining.

What If the Robots . . . ?

The debate about the relationship between machines and employment in the industrialized countries turns on quite different issues in the Third World. Leontief was so conscious of this that he distinguished sharply between the industrialized countries, for which "the specter of involuntary technological unemployment seems to remain no more than a specter," and the developing countries, which are "still waiting in line" to benefit from the waves of technical innovation.[10] The experience of the industrialized countries, in this area even more than others, cannot be transposed to the developing countries. At the macroeconomic level of countries, the compensatory effects are limited to the industrialized world.

The alternative is to choose a far longer period—to think in terms of several centuries before the developing countries can hope to copy the same process of industrialization. This is what Leontief was suggesting, without great illusions, when he wrote that if future improvements produce the same benefits as those of the last 200 years, the developing countries too may hope to advance, "provided their governments can succeed in reducing their high rate of population growth and desist from interfering with the budding of the spirit of free private enterprise." The impact of information technologies on employment then depends on whether the country concerned has a strongly expanding service sector or whether primary products remain all-important.

It is inevitable that relatively fewer jobs would be created as a result of the spread of information technologies in the developing countries than in the industrialized countries. But at the same time, the numbers of jobs lost could well be higher: The comparative advantage of the developing countries lies in lower costs, because of lower wages and the lack of social security coverage, and this advantage is threatened by rising productivity in the industrialized countries. In fact, thanks to technical progress, the industrialized countries are increasingly well placed to produce more cheaply themselves those goods they used to have manufactured in the developing countries. The much-admired skill of young Asian workers was unrivaled in the West as long as microprocessors and computers were hand-built, but now human beings are outstripped by robots, which are essential to produce ever more complex microprocessors in fully automated factories.

For most countries of the Third Worlds, where industrialization has barely even begun, automation in the industrialized countries can have only negative implications. The limited employment provided by the primary sectors is likely to be reduced even further, yet there is little chance of the service sector expanding in the foreseeable future. And one may wonder whether swifter diffusion of information technologies in the developing countries might not occur at the expense of other sectors, and whether this is really the best way of solving problems of employment.

The first obstacle facing the economies of Third World countries is that trade and industry are unable to make up for the jobs abandoned in agriculture because of the irresistible attraction of the towns and cities—one of the characteristic features of underdevelopment is the rural exodus and the massive growth of urban areas without any matching growth in services. Where services do expand, they benefit only a tiny fraction of the population. Thus, the jobs created by the new technical system are generated against a background of nonemployment.

Milton Santos's remarks about urban transport systems in developing countries are even more valid when applied to information technologies. As the sector expands, underemployment and unemployment increase; as one part of the economy is modernized, poverty and technical backwardness worsen for the vast majority of the population. The two parts of the economy are so disparate, the supply of jobs so limited, "space so ill-shared," that the effects of technical change cannot be understood in terms of compensation. The majority at best lives off irregular work in the "lower circuit" of the economy, while the minority with the necessary technical skills has access to the "upper circuit," and the two groups move in completely separate worlds.[11]

There is another reason to have reservations about the possibility of compensatory effects. After all, the industrialized countries themselves do not know when—or above all, if—the gains will ultimately offset the losses. The experience of the past and economic theory notwithstanding, some experts have their doubts on the matter. Hitherto, over the long term, the results were positive, but will this continue? The experts' pessimism arises from the combination of two features of the rise of the new technical system: The information technologies are spreading at a time in the industrialized countries' history when output in the primary sectors is growing without any concomitant growth of employment; meanwhile, developments in microelectronics and the increasing computerization of management tasks mean that productivity in the service sector could improve so much that fewer jobs are ultimately created than are destroyed.

This scenario was followed by Leontief in the book he wrote after his article in *Scientific American*, but this time he extended the scenario beyond Austria and applied the same input-output analysis to the United States through the end of the century. According to his study, which is one of the most exhaustive to date, a high level of automation through the year 2000 could enable savings of roughly 12 percent of the labor force that would have been required to produce the same quantity of industrial output with currently available technologies. The impact on the service sector would be even greater than in manufacturing, as office tasks are computerized—750,000 managers and 5 million office workers in the United States could be out of work by the end of the century.[12]

Nevertheless, despite the enormous potential savings in labor, Leontief concluded that the outcome would not be catastrophic, thanks to the compensatory effects, *provided* that the authorities take the necessary steps to retrain the workforce and redistribute incomes. Under these conditions, displacement would involve many significant changes in the pattern of employment, depending on the sector, occupations, and skills, but there would not be a major increase in unemployment. Everything would depend, once again, on the speed with which information technologies spread. Leontief thought they were unlikely to expand more rapidly because of high costs and social impediments. "The major industrial revolution that began with the introduction of mechanization continued to transform Western economies and societies over a period of about 200 years. The computer revolution became apparent only a few years ago, and it will be no further advanced by the year 2000 than mechanization was, say, around 1820."[13]

The compensatory effects may operate in favor of the industrialized countries, but it is hard to see how they will help solve the problems of underemployment in Third World countries. On the contrary, Leontief's scenario suggests that their position of labor surplus can only worsen, as a result of population growth and the industrialized countries' success in adjusting. Leontief's study did not look beyond the end of the century—another fifty years and the Information Revolution could have extended the savings in labor to the whole planet.

In fact, there has never been any equivalent to the introduction of robots in the whole history of mechanization. Unlike other technologies that raise productivity, the robot *replaces* the worker. This latest stage in the Industrial Revolution combines design, manufacturing, and marketing in a single stream of information, so that we could if need be automate everything we do not want to do ourselves.[14] During the 1950s, the unions in the United States were worried about the risks of unemployment arising from automation; the debate fizzled out because the technologies were not ready and high rates of growth led to full employment. But the reasoning that was reassuring in the past could prove less and less plausible in the future.

Whereas once robots were simply able to copy certain human actions, they can now supplant the human worker. The idea of fully automated factories is no longer utopian, and robots can in fact make other robots. Advances in artificial intelligence are permitting an ever greater mastery of automatic functioning and hence better integration of varied actions and tasks. The industrialized countries see this process as leading inevitably to the adoption of social and political measures relating to working hours, work sharing, and income distribution. It will alter not only the structure and content of employment but also the institutions, organization, and values of economic life. Technical progress will transform the very nature of work and leisure, creating jobs and activities that are less and less like traditional production tasks.

The higher the standard of living in a country, the smaller the proportion of the workforce employed to produce the goods and services required to satisfy the basic needs of the population, and the greater the time devoted to leisure. Work is not an end in itself—it only appears to be because rigidities in the social system block the necessary adjustments in terms of different working hours, distribution of income, and so on. It is only when the imbalances are thought to be unavoidable that this constitutes a high-priority challenge to be tackled. But if the necessary adjustment can be made, what becomes an end in itself is the relationship between production and aspirations other than work.

The history of mechanization is also that of shorter working hours: The real question is whether robots will put people out of work or send them off on vacation. Too many economists, fascinated by the short-term effects of technical change, see only the job losses, when they should also be considering the gains in leisure time and the continuing opportunities of adjustment. That idea lies behind Leontief's parable:

> Adam and Eve enjoyed, before they were expelled from Paradise, a high standard of living without working. After their expulsion, they and their successors were condemned to eke out a miserable existence, working from dawn to dusk. The history of technological progress over the past 200 years is essentially the story of the human species working its way slowly and steadily back to Paradise. What would happen, however, if we suddenly found ourselves in it? With all the goods and services provided without work, no one would be gainfully employed. Being

unemployed means receiving no wages. As a result, until appropriate new income policies were formulated to fit the changed technological conditions, everyone would starve in Paradise.[15]

The developing countries find themselves, by definition, in the opposite situation. Their problem is not to reduce the work time needed for productive purposes but to expand employment and output simultaneously. From this standpoint, the prospects opened up by technical progress are far from reassuring, especially in view of demographic trends. The Information Revolution cannot play a central part in solving their problems; on the contrary, it threatens to make the problems even worse.

Unless one supposes that the rich countries, as they work less, are able and willing to help the poor countries to the point of taking total care of them, there is no alternative but for the poor countries to raise their own income for their own development. Where will the jobs, the surpluses, and the resources come from when machines in the industrialized countries take over increasingly from the human labor force in the underdeveloped countries? It is for the countries of the South, far more than for the North, that robots threaten to be a revolutionary and victorious rival. "The worst does not always happen" was the subtitle of Paul Claudel's *Silk Slipper,* in which he celebrated in his own way the transformation by European faith of the lands that are now the developing countries.

8 The Cathedrals in the Desert

Knowledge is power: Bacon's dictum does not apply simply to those matters whereby the scientific revolution of the seventeenth century gave Europe access to a whole new world; it still marks the divide between the scientific culture and practice of the industrialized countries and that of most developing countries today. All civilizations—ancient, modern, primitive, traditional—operate on the basis of a concept of knowledge and links between that knowledge and action. No other, besides the one first created by the scientific revolution in Europe, has turned theoretical knowledge into a lever to achieve practical ends. The application of mathematics to natural phenomena and the experimental method together made systematic prediction possible; this was then carried out rigorously, and the knowledge used to act upon and transform the world. The tool is no longer something *given,* at best the product of practical know-how, but is deliberately *constructed,* the product of an alliance between science and technology.

The scientific revolution was not merely the substitution of one model of knowledge for another, of mathematical and experimental knowledge replacing perception by the senses. It also involved a worldview holding that the ability to change things is directly dependent upon speculative knowledge: Theoretical knowledge is intended to be useful. In order to illustrate the gulf separating this view from all others, historians of science make the contrast with Greek or Roman science, though one could just as easily refer to Chinese or Indian traditional science, which is still widely found in Asia.

To take one example, historian of science and expert on Indian medicine Francis Zimmerman has shown why the worldview of traditional India could not have given rise to natural science as we understand the term. The principle behind ayurvedic science is one of law: "Beings are classed according to their merits and their ability or inability to perform the rites. The idea of a *science* of Nature is completely alien to India . . .

or to be more precise, the idea is formulated in an utterly different way."[1] In this scheme of things, medicine does not involve intellectual understanding, but has to do instead with rites or myths, and medical acts derive from methods laid down in the sacred texts and not from an attempt to discover causes. *"Everything useful is given,* either perceived or through Revelation and Tradition which complement perception."[2]

In short, knowledge comes from outside, based on perception or oral transmission and texts; there are no experiments, and tradition is paramount. This does not mean that such medicine has no beneficial effects, but its efficacy cannot be explained rationally, and thus the principle underlying its results cannot be applied universally. That, however, is not its purpose, whereas the very essence of Western science is the universality of its operational capacities: Searching for and identifying causes underpin an efficient response to the effects.

Pasteurian medicine is thus in the direct line of descent from the scientific revolution of the seventeenth century because it conducts experiments on the invisible, discovering the causes of illness in things that cannot be perceived: genes, bacteria, viruses. Its clinical efficacy derives from its ability to grasp intellectually what is wrong. Accordingly, as Claire Salomon-Bayet has argued, the best place for its research into disease is not so much the hospital—"the sickbed of the sufferer of the neo-Hippocratic tradition"—as the laboratory. Treatment involves "an effort to understand what constitutes the organism or the micro-organism, the human being or the cell, *living things* as the object of science."[3] The experimental approach results in universally effective vaccines; knowledge gained in the laboratory results in the medical uses of microbiology and synthetic drugs.

Knowledge is power: The dictum is even more valid for the information technologies than for all those in the past based on matter and energy, because these new technologies mechanize functions hitherto the province of the human mind by accelerating, increasing, and broadening its scope to a prodigious extent. There are, indeed, direct consequences for power in the political and military sense—as the Gulf war showed, in modern warfare involving very sophisticated weapons, however "conventional," the strength of what is termed "C cubed" (systems of command, communication, and control) would count for far more than the physical might of the arms themselves.

As a Thundercloud Brings the Storm . . .

The additional knowledge contributed by the new technologies amounts to more than just tools for handling and transforming matter and energy. The Information Revolution goes beyond the technologies and involves a radical change in society, perhaps of our whole civilization, affecting attitudes of mind, behavior, and values. This new knowledge is all the more powerful because it concerns the collective intelligence: language, memory, culture, organization.

In the industrialized countries, there are worries about the unfair distribution of opportunities offered by the new technologies, with fears that a growing proportion of the labor force will be condemned to unpleasant, low-paid jobs without hope of social advancement. This unfair distribution is (and will be) even worse, inevitably, in the developing countries. The risk that the proportion of the population unsuited to cope with the Information Revolution will increase rather than decrease is even greater. The gap can only widen between the technically skilled groups able to make the most of the better working conditions and the advantages brought by electronics, and the vast mass of those left behind, at best toiling away at the arduous tasks of a past era.

Threats of such sociocultural rifts apply not just internally, in both industrialized and developing countries, but also internationally, between rich and poor countries. For one thing, transfers of information are in fact extremely costly. Computers and telecommunications provide the most efficient method ever invented of circulating information and making it available to all, but information is also a marketable commodity: Thanks to the new technologies, certain types of information that used to be free are now expensive. Memories relating to traditional craft methods, for example, have hitherto been freely available, but now they are being replaced by computerized systems that are costly and difficult to access. For developing countries, the costs may be prohibitive, and barred from access, they will become ever more blatantly handicapped as regards knowledge.

Computers and telecommunications already allow instant access in theory—and increasingly will do so in practice—anywhere on earth to the global stock of organized knowledge. A veritable cultural revolution is taking place, very similar to the one brought about by the spread of printing. After all, the transition from the written to the printed word was not simply responsible for popularizing reading; it provided science with new means of standardization, of storage, of dissemination, and hence of discussion and research based on shared data.

This is the theme running through Elisabeth Eisenstein's study of the history of printing: "One cannot treat printing as just one among several elements in a complex causal nexus, for the communications shift transformed the nature of the causal nexus itself. It is of special historical significance because it produced fundamental alterations in prevailing patterns of continuity and change."[4] Gutenberg's technical invention thus directly encouraged advances in theoretical knowledge by spreading and stimulating the intellectual revolution and enabling more and more specialists to read "the book of nature." This was not simply thanks to reliance on mathematics rather than revealed scripture but because copies were now free from scribal errors and hence could be studied on equal terms everywhere.

Information technologies are the source of equally spectacular and radical changes—though these occur far more rapidly—in the way scientific research is conducted and the results used outside the scientific community. They are both cause and effect of the globalization of trade, commerce, industry, stock markets, and press agencies. Data banks may make information, even from distant sources (notably the world's best libraries), more mobile and accessible anywhere on earth, but they are far more expensive than newspapers, journals, and books. In this area it is difficult to separate the carrier from the carried, the medium from the message: Their contents are far from unbiased.

The supremacy of U.S. firms in the area of bibliographical data bases and their dominant position with regard to data banks carry a risk of intellectual alienation that the industrialized countries are the first to criticize, lest the reorganization of knowledge around this new method of classification, formulation, and re-creation of the collective memory lead to a loss of identity. The vision of the world that the *New York Times* data bank presents is not the one nations other than the United States may have of themselves. But if these countries are obliged to seek their sources from that data bank, what image are they going to have of their own history? The computer as the "destroyer of history" is not a science-fiction nightmare; as early as 1976, a computer expert at MIT, Joseph Weinzenbaum, called attention to the risks of new forms of cultural dependence arising from transborder data flows.[5]

This dependence seems inevitable for the developing countries that cannot set up their own data banks and must rely on foreign firms and memories. The pool of technical information (agricultural, medical, socioeconomic, cultural) on which the developing countries must draw is based overwhelmingly on the experiences and the methods developed in the context of the most advanced industrialized nations. Quite apart from the fact that 95 percent of scientific and technical information is now held solely by the industrialized countries (two-thirds of it in the United States), many developing countries have to purchase abroad information pertaining to their own national circumstances. Dispossessed of their history, they even have to pay to gain access to it.

Information technologies are the product of a culture that may acquire global domination at the price of cultural impoverishment for many countries. Computers so far have depended on the structure of language—in particular, software written in English, which some people think has an increasing impact on modes of thought, economic and social organization, and cultural behavior. Research in the major computerized data bases, at any rate, imposes Western concepts, which may be totally alien to users brought up in a different language and culture.

The debate about the threat to cultural identity posed by information technologies is not a new one: It started and grew with radio, the cinema, and most especially television. Microelectronics—less costly and more flexible and versatile—could lead to a diminution of cultural dependence, provided that more countries were better placed to devise and produce the software they need. However, the link between computers and telecommunications gives new impetus to the debate as satellites now permit the instantaneous direct transmission of visual images, a development that promises to make an even greater impact because of its immediacy. There is also a cost involved, because countries that cannot create their own independent broadcasts, films, and cassettes are bombarded daily with the products of foreign news networks.

Transborder data flows are growing at the rate of 15 to 20 percent per year, but that figure indicates only the volume of the transactions and not their content and nature. The last, obviously, is the most critical. No customs control can be imposed without interfering with the privacy that protects all correspondence: Magnetic tapes can be taxed, but not the information they contain. Furthermore, the new technologies directly threaten the public monopoly over postal services and telecommunications. For example, the reduction in the cost of satellites and especially of the land-based receivers linked to them is leading to a proliferation of private television networks that can saturate whole regions without regard to local sovereignty.

The technical battles have immediate political dimensions, as can be seen from the discussions within the International Telecommunications Union (ITU) on the sharing of wavelengths. The countries of the South complain that the North benefited from the principle of "first come, first served"—that because the most easily accessible (and hence least expensive) wavelengths are saturated, they are now obliged to seek places on ever higher frequencies. To do this requires access to the latest technologies, which are beyond the reach of the developing countries. The same applies to orbiting satellite stations, which provide the best possibility for developing countries to establish and improve their domestic telecommunications networks; the best positions are already taken.

The cultural issues arising from the information technologies cannot be separated from the political and economic ones. The old saying that "The person with the information has the power" lies behind the tensions generated by NOMIC (acronym of a group of nonaligned countries seeking a "new world order" with regard to information). Since its formation in the mid-1970s, NOMIC's program has created a dialogue of the deaf between North and South at one United Nations conference after another. The South demands a fairer distribution of information channels so as to avoid domination by the North, but the problem is to separate the contents of the message from the ownership of the means by which it is communicated.

All the international press agencies belong to the industrialized countries: The biggest are U.S. (Associated Press), British (Reuters), French (Agence France Presse), Japanese (Kyodo), and Russian (Tass). How can better balance be achieved? The world of intangibles has fuzzy frontiers that allow ideology and good intentions to coexist easily. Where the industrialized countries talk about freedom of information and free trade, the developing countries see domination and neocolonialism; where the industrialized countries stress freedom of the press and opening up markets to assure the transfer of information like any other commodity, the developing countries demand state intervention and protectionist policies in order to guarantee a fairer sharing of the networks and their contents.

The way to an information hell is, like the others, paved with good intentions. NOMIC's program not only means denouncing, behind the "distorting mirrors" of the international media, "the colonization of minds," but it also encourages an interpretation of the role of the state with regard to information and to the protection of journalists that offers a certificate of respectability to all sorts of dictatorships. In these struggles, words can easily be understood as their opposites—"access," "participation," and "sharing" suggest a concern for the principles of democracy and unchallenged freedom of the press, but if they mean that state intervention should govern the way the media function, there is every reason to be wary.

> Any domination, even "just" (if one can say that) economic or technological, carries with it cultural imperialism as a thunder cloud brings the storm. Great vigilance is necessary to prevent technical and cultural cooperation, or imports of equipment and technologies, from imposing an inappropriate model of development, encouraging a brain drain and, in the guise of a valuable cooperation, leading to a new form of dependency.[6]

That caution is well put, and in a Leninist style that prompts the question, What is to be done? The negotiations on the NOMIC proposals are condemned, by definition, to ever more empty statements, without any practical follow-up.

There can be no illusions about the chances most Third World countries have of becoming independent producers rather than consumers of information. The technological domination of the North clearly requires adjustments in favor of the South, but the decisions taken must not create the opposite of what is wanted: in the name of sharing information, to give control over it to the state. There are the same traps inherent in the right to information as in the transfer of any technology. It is not enough just to make the information available; there must be the means of using it properly.

The term "information" covers a range of activities—commercial, technical, military, diplomatic—but at the end of the transfer process, it includes above all a collection of knowledge on which depends mastery of the physical networks that embody the technological hegemony of the North. Is it not rather cultural domination that carries with it, like the storm cloud, economic and technological imperialism?

Technology Transfer and Transport

Importing production techniques designed and put into operation by the industrialized countries has three drawbacks for the developing countries: They get a technology that is often ill-suited to locally available resources, in particular the abundant supply of unemployed labor; overcapacity in relation to the size of the local market; and a very small gain in productivity across industry and trade as a whole. For example, industrialization in the developing countries has generated a relatively small amount of employment. It is true that Argentina, Brazil, and Mexico (which together account for 42 percent of the total manufacturing output of the Third World) have sometimes equaled or even surpassed the ratio of industrial output to GDP of the industrialized countries (37 percent), but the ratio of employment in manufacturing to the total labor force has remained far lower (22–29 percent against 39 percent).

The dominant role of multinational corporations in the process of technology transfer has been pointed out so often that one hesitates to raise the subject again. Nevertheless, it is important to recall the fact that most of the transfers do not occur via direct investment but through internal flows within multinational firms. Hence arise the criticisms directed at these companies: The developing countries are at their mercy for products and processes for which the costs have already been recouped, whereas access to the know-how that would give the developing countries opportunities to produce independently is forbidden.

When the multinationals set up laboratories in developing countries, the purpose is not to conduct research into the latest and most efficient technologies for world markets but to adapt and exploit medium- or low-technology products for the local market. By contrast, in the 1970s, in order to benefit from a less well-paid workforce with lower social security costs, these firms opened more and more factories throughout the Third Worlds to produce the most sophisticated products for sale in world markets, a move taken because relocating research is less profitable than relocating production—and even that becomes a less attractive proposition when fully automated factories make it possible to conquer new markets thanks to massive output of innovations rather than to low prices.

Decisions taken outside the economic, social, cultural, and political space of the developing countries determine when the new technologies emerge, are put into operation, and spread. The risks of upheaval are made all the greater insofar as these countries lack the means themselves to assess, choose, assimilate, and adapt foreign technologies. From this standpoint—and this is what makes studies of them so ambivalent—the multinationals are both a factor promoting faster industrial transformation and a threat to the existing balance among manufacturing sectors, if not to the entire economies, of the receiving countries.[7]

Mastery of a technical system means not merely being able to use it but also knowing how to make repairs and hence handle the various components. It is not enough just to move machines and their instruction manuals for them to function in the same way anywhere in the world. As long as a country lacks the capacity locally to deal with a technology, then the technology has been *transported* and not *transferred*. In economic terminology, one might say that the technology has simply been supplied in response to a demand. Nonetheless, technology supply is never so straightforward as to require no more than an instruction manual, nor is the demand phrased so clearly as to be a package of needs stated once for all.

The history of major development projects is full of examples of technical systems that failed to work because the wrong things were supplied, because neither the conditions nor the prices were those agreed, and above all because of a lack of qualified staff to maintain and run the plant. There are also many instances of projects for which both the potential of the equipment and the real needs of the local market were overestimated, so that they became white elephants, with no purpose, no customers, and no future.

Some of these projects were urgent only in relation to the ambitions of the ruling heads of state, and they became useless and costly examples of what economists call "demonstration effects" and others label more bluntly delusions of grandeur. The weakness of the rulers of many developing countries for prestige projects gives some firms in the industrialized countries extremely attractive and lucrative market openings, with the added bonus that the massive investments entail no risks because the projects are usually financed largely by the home state of the firm involved through "tied aid" rather than by the developing countries themselves.

These white elephants abound, the result of collusion between the rich countries' wish to make money and the poor countries' wish for magnificent display. They are half comic, half absurd, according to Stephen Hill, an Australian with more than twenty years' experience in development assistance programs in Asia. They include unused nuclear power stations, which were sold on condition that the receiving country agreed to take the radioactive waste of the supplying country; an aircraft bought for the private use of the government's leader, but which was so big that the runways of the local airports, built on coral reefs, had to be extended at vast cost; steam turbines intended to be fueled by eucalyptus wood, on the suggestion of a U.S. expert who clearly did not realize that trees would have to be planted and would not be ready for six or seven years, so that in the meantime fuel is being found by tearing up all the coconut palms in the country; or the ultramodern hospital whose annual operating costs, notably for electricity, are greater than the capital costs of its construction; and so on.[8]

The bigger the development project, the more experts, consultancies, and firms in the industrialized countries become involved, without necessarily taking account of the interests of the recipient country. Mickès Coutouzis based his doctoral thesis on a project of the International Energy Agency (IEA), which was created at OECD at the time of the energy crisis. He followed the various incarnations of this project through its ill-fated history. The account is especially revealing because it gives a "fly on the wall" view of the negotiations, shifting alliances and interests that go into technology transfer. He tells the story of a Greek urban planner, now a U.S. citizen, who specialized in systems analysis and who designed a mathematical model of an energy-saving town. He proposed that the IEA should test his model by creating a new town of 3,000 to 5,000 inhabitants on the island of Crete, where all services would run on solar energy. The U.S. Department of Energy gave its backing, the IEA launched the project, and the Greek authorities took part in the study mission, which selected a site where there were two or three villages with about 750 inhabitants and some goats.[9]

The project involved years of work, the expenditure of millions of dollars, and constant interaction among its many facets—calculations by the consultants, negotiations with the authorities, surveys and public participation initiatives among the local population, contributions from academic experts, and political intervention from all sides. The general methodology designed in the United States was meant in principle to be replicable anywhere, and the new town to be a prototype community, providing a model that could be exported to all the sunny regions of the world. But the realities proved resistant to the mathematical fantasies of the planners. The local residents wondered why everyone was suddenly so interested in them (the rumor quickly spread that there was actually a plan to build a U.S. army base), and the experts from Europe and the United States wondered where they would find the 3,000–5,000 people who were meant to live in the prototype community.

After a process that had mobilized the interests of everyone except the locals most concerned, the chief administrator of the region stood the whole problem on its head and put all the mathematicians, planners, systems analysts, energy consultants, and promoters to flight by saying that what the region really needed was not improved energy supplies, however economical, but rather water. The only way for the region to develop would be if there were a program to provide a less expensive system of irrigation, which the inhabitants could use as they saw fit; they did not need a model community that the experts had not thought to try out in a more densely populated area, perhaps one more well-disposed to the benefits of technology and modern life and in any case with a larger population, more jobs, and resources—for example, in California.

In the draft plans of the consultancy firms and the systems analysts, the prototype community was a theoretical concept that had a life of its own. Its designers thought it could be set up without modifications, regardless of local conditions. Of course, as soon as the irrigation scheme parted company with the model of an all-solar town, both the U.S. Department of Energy and the international organizations ceased to be interested, and the experts abandoned the locals to their goats. The transfer of a universally replicable technical solution ended up as a utopian exercise, and specific local characteristics won the day. A technology does not operate as "a separate and independent whole, with society *beginning* outside it, and then shifting it, unchanged, from seller to user." Societies and technologies move together or not at all.[10]

Technology transfer between industrialized and developing countries involves a social process that cannot be reduced to the straightforward shift of a technology from one place to another. First of all, there are legal issues relating to the right to use methods developed somewhere outside the receiving country. Sales of capital goods and whole plants are just the most visible type of technology transfer, which also includes sales of patents, licensing, and the provision of services.

China, with no tradition of written law, came up against this aspect when it instituted the policy of opening up the economy: The Chinese discovered that they needed patent laws and commercial law in general if they wanted to attract Western firms to take part in joint ventures. Concern to reassure foreign investors—and at the same time to stop these firms from curbing the export of products manufactured under license—brought a new legal system swiftly into existence. But it was not so simple to put it into operation, partly for historical reasons. Foreign firms would sign a general agreement and then find themselves dealing with obscure regulations they had never heard of, because the tradition of imperial law (which depended on the good will of an omnipotent and divine emperor) combined with a host of regulations at the provincial, local, branch, or even firm level, as well as the influence of Soviet law, to offer opportunities for interpretations of every kind. In addition, the Chinese were completely unfamiliar with the legal methods current in the industrialized West. As a result, Chinese teams were sent to study patent laws in the United States.

The right to use technologies in the strict sense cannot be separated from the right to use the prerequisite knowledge and hence from the right of access to the information that must be mastered if the imported goods or processes are to operate satisfactorily. Part of this information is set out in diagrams, descriptions, and advice. An equally large part cannot be written down, because it depends on experience built up over time and learning by trial and error; needed is the know-how, individual or collective, of those who have learned to regulate the machinery, manage the production procedures, and resolve problems of running, maintaining, and handling imported machines.

The physical technology is just the visible part of the iceberg. The hidden part retains all its secrets unless they are deliberately passed on, which occurs in ways that are more reminiscent of apprenticeships in medieval guilds than of the methods of the electronics age. This is even more the case when factories are sold as turnkey projects: The transfer of capital goods means that engineers, technicians, and supervisors must also be brought in to make the factory work and to train the local workforce.

The process of transfer does not, however, stop there. When the factory is in operation and the products are ready to be sold, perhaps abroad, local people must learn the techniques of marketing. The consultancy firms must then expand their advice from construction to deal with knowledge transfers relating to work organization, market research, banking practices, and the like. In this context, it is never the technology by itself that matters so much as the associated know-how and the whole range of supplementary knowledge and methods dealing with production, management of services, marketing, distribution, and sales.

Weapons Versus Development

Technology is a dead letter if no attention is paid either to the training required to make the imported item work or to the context in which the output is to be sold. In a 1986 article in *Revue Tiers Monde,* Claude Courlet and Pierre Judet refer to a study of 343 industrial plants built in sub-Saharan Africa over the previous twenty years showing "that 274 function badly or not at all (79 are no longer running), and about 60 are using their productive capacities properly." The authors show that the most common reason for malfunctioning relates to a lack of any real tradition of manufacturing as well as not enough small and medium-sized firms locally.[11]

The authors point out that the developing countries are not the only ones to have "cathedrals in the desert," or planning mistakes. It is true that the Fos complex in France (a huge industrial plant built around oil refineries) or the combined steel and chemical plants in southern Italy, built at vast cost, did not become "growth poles" as it was hoped. But the long-term costs of these errors do not bring the same penalty in the industrialized countries as in the developing countries. For the former, these misjudgments are small blots on a very diverse industrial landscape, whereas for the latter, the cathedrals are all they have as industrial landscape. At worst, policies of restructuring force the industrialized countries into redeployment, and they have the resources to make good (more or less) any structural imbalances. For the developing countries, these cathedrals stand out as monuments to short-lived ostentation against the background of lasting low productivity of the whole economy.

For these reasons, one may wonder about the future of some of the major industrial projects built in the Arab countries with their oil revenues. Since the 1970s, industry in these nations has grown strongly (8–10 percent per year) from a very low base, but these results are only superficially impressive because in most cases manufacturing still accounts for less than 10 percent of GDP. The huge investments encouraged cost overruns; the oil wealth admittedly made possible large-scale purchases of food, but the fact that food is massively subsidized by the states (amounting to 40 percent of the Egyptian budget) has merely added to the problems of the farming policies and sometimes contributed to the decline of agriculture.

Where production is geared to world markets (as in the case of Saudi Arabian petrochemicals or the aluminum and fertilizer plants in Bahrain and Qatar), substantial export incentives are given. In the medium term, the reserves of oil and gas will not be able to compensate for the deficit in manufacturing, unless oil prices and the dollar exchange rate rise. In Algeria and Egypt, as in Saudi Arabia or Syria (it does not matter whether the regimes are socialist or capitalist), the public sector acts as welfare state and state entrepreneur, and this results in the inefficiency of the productive system and the paucity of small and medium-sized firms. Most of the population in fact receives assistance, and the local firms cannot cope with the growing demand for consumer goods.[12]

Without subsidies, the large firms cannot supply goods at prices competitive on world markets, and the small and medium-sized firms cannot satisfy local needs. Manufacturing has few linkages upstream or downstream: Not only is it isolated from the traditional craft sector, but it also has difficulties in diversifying beyond petroleum byproducts, so that the whole economy is increasingly vulnerable to the ups and downs of the oil price. Having failed to coordinate their industrialization programs (instead competing with each other in the same limited sectors) and having failed also to encourage the growth of small and medium-sized firms, the Arab states are now at risk of finding themselves unable to sustain the costs of their commitments to the welfare programs that the manna of their oil revenues allowed them to overlook in the 1970s.

Some of the oil producers, such as Saudi Arabia, must quickly reduce the immigrant labor force that the big projects attracted. The Gulf emirates are finding that they must cut back on the services they used to provide free of charge and introduce direct taxation. Others are burdened by large numbers of public employees in administration and state-owned firms (e.g., Algeria and Egypt) and face revolts when subsidies on food are cut (bread riots). The very few examples of diversification or occasional successes cannot hide the precariousness of social tranquility, further threatened by the attraction of fundamentalism. The period of great industrial ambitions is over, leaving some of the major projects unfinished, and oil revenues risk becoming a water bag riddled with holes. Nothing prevents the technological cathedrals from ending up like the Pyramids, abandoned in the desert, their contents plundered.

One of these follies is especially unproductive and constitutes a greater burden on the developing countries than all the others: arms purchases. In the shadow of the atomic bomb, the industrialized countries have avoided direct confrontations since 1945, but they have given their direct or indirect support to many regional and local conflicts in the developing countries, to the point that the latter suspect the North of maintaining "nuclear peace" at home while dumping their quarrels on the developing countries' backs. Nevertheless, of the 200 or so armed conflicts since the end of World War II, more than half have been internal troubles, revolts against the ruling regime, or tribal, religious, or separatist wars.

It is true that the Northern superpowers have gone on talking about arms control while they continued to update their stock of nuclear weapons and intercontinental missiles. But the countries of the South talk about development, and all the while their defense expenditures have been growing faster than those of the industrialized countries. Furthermore, the costs of maintaining their police forces have risen more rapidly than expenditures on classic defense: by 144 percent in Africa between 1966 and 1975, whereas the budgets for conventional armed forces rose by 40 percent.[13]

In any case, the arms trade has meant a staggering increase in technology transfer. Twenty years ago, the newly independent nations were satisfied with cut-price or secondhand weapons, whereas now all the poor countries covet the latest models to come off the assembly lines. The most sophisticated weapons for conventional warfare are as much a status symbol as a means of defense and security. The breathtaking rise in the price of modern military materiel seems not to deter most developing countries from placing orders. There is no more striking contrast than that between the success in transferring military technologies and the fall in incomes or rise in debt of the Third Worlds.

The arms trade is a very profitable business for the selling countries, which has encouraged a handful of developing countries to start up independent arms industries and to compete in the markets hitherto monopolized by the industrialized countries. In the early 1970s, twenty-seven developing countries manufactured some proportion of the equipment needed by their armed forces, but this was only guns and munitions. Since then, eleven have started aircraft industries, six produce missiles or armored vehicles, and nine have a shipbuilding capacity. It is estimated that the value of the arms and equipment produced in the Third World (China excluded) rose from less than $1 billion in 1970 to more than $5 billion in 1980. These figures are admittedly absurdly low compared with total expenditures on armaments (over $600 billion), so that the developing countries account for barely 1 percent, and most of those weapons are made under license.

In fact, the capacity to produce arms (including the R&D side) remains in the hands of a very few developing countries, with Brazil, China, and India accounting for the lion's share. These are countries whose industrial potential has been strengthened in the last twenty years in conjunction with their ambitions to have independent arms production. Brazil, for instance, has about 350 firms, employing more than 100,000 technicians and manual workers. The best known are the state-owned IMBEL and EMBRAER and the private firm of Bernadini. Some are well established in the export business, with a catalog of products whose price and performance are very attractive for customers in Africa and the Middle East (e.g., the Xavante jet trainer, the Cascavel armored car). These products are all the more attractive insofar as they allow the purchasers to escape from the state of dependency normally imposed by contracts with the most industrialized countries.

The growth of the arms trade in the Third Worlds and the emergence of new producers such as Brazil, Argentina, South Korea, and Taiwan are a sign both of the success of the industrialization policies being implemented by some of these countries and of the sense of insecurity rightly or wrongly obsessing most of the others. The surpluses created by one group in order to produce and by the other in order to buy do not thereby contribute to development. The displacement of armed conflict to the poor countries benefits only the arms dealers, and the militarization of the Third Worlds reinforces the uncertainties of an already precarious economic growth.

In the 1970s, theories proliferated about how to loosen the vicelike grip of technological dependency.[14] But this dependence is now stronger and more inevitable than ever, and the developing countries themselves make the stranglehold worse, particularly when they choose to get involved in vast projects for reasons of prestige or in the arms race. The theories on dependence come down to this: To escape from the distortions caused by trade imbalances, the dominance of the multinational corporations, and the model of the consumer society provided by the industrialized countries, it would be best to cut loose from the international economy—to avoid all technology transfer. But technology transfer, too, is unavoidable unless the country in question is willing to turn its back not only on the rest of the world but also on all chances of development.

The real question is not whether the developing countries can escape from technological dependence, but why it is that some countries manage to use their dependence to construct a relatively independent growth path, and at what cost. Obviously, the mastery of use and even more the mastery of production of the new technologies presuppose that conditions are satisfied that not all developing countries can hope to meet. The imbalance of technological exchange is also connected with the unequal distribution of opportunities in what David Landes has called the chase to industrialization: "a race without a finishing line [in which] there are only a few contestants sufficiently endowed to vie for the palm."[15] There is no evidence that the example of the newly industrialized countries can be replicated or indeed that development occurs only thanks to the most advanced technologies and machinery.

9 The Newly Industrialized Countries

Joseph Needham deliberately chose to end his monumental nine-volume history of science in Chinese civilization at 1601, the year the Italian Jesuit Matteo Ricci, who had at last received permission to live in Peking, filled the court (and especially the emperor) with wonder at his scientific knowledge and then began to initiate learned Chinese into the experimental method. That year was symbolic as a turning point. According to Needham, it was the moment when Chinese science started to blend in with European science.[1]

Matteo Ricci acted as a technical assistance expert, no more and no less disinterested than all those who would follow him in the nuclear and electronic age. Europe was looking for new souls to convert and new markets to conquer: The missionaries opened the way, as we would now say, to expanding trade but also to bringing the gospel to as many Chinese as possible. Ricci gave courses in mathematics and astronomy, translated Euclid's *Elements* into Chinese, helped reform the calendar, and made known such items of Western technology as solar clocks, a new type of mapmaking, and cannon foundries. At the same time, he translated the Ten Commandments and the catechism, as well as many works of Christian apologetics, and he preached the gospel. Western technology traveled in company with religion.

Nevertheless, neither caught on for quite a long time. The Jesuit's Dominican and Franciscan colleagues blamed him for going too far in his campaign of evangelism. He adapted Christianity to Chinese beliefs and declared that Confucianism and ancestor worship were not incompatible with worship of the Virgin Mary. He realized that Christianity had no future in China unless it was mingled with traditional beliefs, and missionaries who disregarded this fact risked expulsion or massacre. When he died in Peking in 1610, honored by the Chinese as a great mandarin, his efforts had already been violently attacked and denounced in Rome.

That was the start of the quarrel among the European missionaries, with the Jesuits' Chinese-style catholicism increasingly appreciated by the emperor and criticized by the other Christian missions. Papal legates were dispatched to China to investigate, criticism mounted in Rome, but the disavowals did not prevent the Jesuits from carrying on, with the support of the emperor K'ang-hi, who went so far as to grant an "edict of tolerance" to the Christians in 1692. The virulence of the attacks and the spectacle of the feuds among the Christians made K'ang-hi feel that his broadmindedness brought poor returns, but he continued nevertheless to protect the missions. His successor Yong-cheng was less cooperative: In 1724, two years after his accession, he banned Christianity, confiscated the churches, and expelled all the missionaries—except the Jesuit mathematicians in Peking.[2]

The Crucial Difference

Ultimately, the failure of Christianity was caused by intolerance, not on the part of the Chinese but of the representatives of the Holy Church. The failure of Western science, by contrast, was the result of the rigidity of traditional Chinese culture and social organization. European science grew very slowly in China in the eighteenth and nineteenth centuries, blocked by the same factors in Chinese society that had always been an obstacle, according to Needham. But at the time when the scientific revolution was starting in Europe, one field was making great strides in China, namely the humanities. In their rigor and precision, the methods used were not unlike those underlying the new phase of European science. Needham quotes Hu Shih, the historian who made a special study of this renaissance:

> Four years before Ku Yen-Wu was born, Galileo had invented his telescope and was using it to revolutionise the science of astronomy, and Kepler was publishing his studies of Mars and his new laws of the movements of the planets. When Ku Yen-Wu worked on his philosophical material and reconstructed his archaic pronunciations, Harvey had published his great work on the circulation of the blood, and Galileo his two great works on astronomy and the new science. Eleven years before Yen Jo-Chhü began his critical study of the *Book of History,* Torricelli had completed his great experiment on the pressure of air. Shortly after, Boyle announced the results of his experiments in chemistry, and formulated the law that bears his name. The year before Ku Yen-Wu completed his epoch-making *Five Books* on philological studies, Newton had worked out his calculus and his analysis of white light. In +1680 Ku wrote his preface to the final texts of his philological works; in +1687, Newton published his *Principia*.[3]

Similar methods, different topics, comments Hu Shih. Progress in philology on one side, progress in natural science on the other. The West worked with stars, levers, slopes, and experimental proofs, while China worked with books, words, and documentary proofs. The Chinese renaissance generated knowledge in the humanities that allowed further study through books; the surge of science in the West created a new world in which theoretical knowledge made it possible to master matter and energy. Needham concludes that nothing prevented the experimental method from emerging in China at that time—except the view of the world and the social structures from which that attitude derived. But this difference was crucial: Neither the Jesuits' mathematics lessons nor their technology transfers could take root in such terrain.

When we wonder why the Industrial Revolution did not occur in China, we are implicitly asking why modern science did not develop there earlier. After all, medieval Europe had acquired much of its technical knowledge and methods from the Far East. The three most important technical innovations of the European Renaissance—the compass, gunpowder, and printing—were all Chinese inventions, and Needham's study indeed shows how far ahead of the Europeans the Chinese were at that time.

The success of the scientific revolution in Europe is inseparable from the emergence of modern capitalism: The former had to do with a combination of behavior, institutions, and values that cannot be dissociated from the orientation given to economic and technical activities. It may be arbitrary to stress one single factor from among all those involved, but it would be difficult to exaggerate the influence of the approach and the practical techniques of the experimental method. The mechanization of industry was as much the result of the beliefs, attitudes, and social institutions that had produced that method as it was of the practical applications that could now be multiplied. To change the world—to "make Man the master and possessor of Nature," as Descartes said—means drawing on a conception of time that does not recognize circularity and repetition.

It is characteristic of traditional societies to be in a state of equilibrium that is upset only by external factors such as wars, invasions, epidemics, or natural catastrophes. Technical transformations happen gradually, taking centuries to occur. The peculiar feature of industrial societies is that, in addition to these external destabilizing factors, they respond to internal pressures that condemn them to constant upheaval and change. Clearly one of the most important of these internal factors is technology, but that does not mean technology alone shapes, let alone determines, economic and social change. There is an interplay of reciprocal influences among the introduction of new technologies and the economic and social structures on which technical change makes its impact and spreads. The response of individuals, groups, and societies is not passive; technical change is the subject and the central issue of negotiation, conflict, and compromises, which alter the nature of the new technologies just as much as they affect human attitudes and behavior.

The experimental method created every aspect of its techniques, which were always geared to achieving a goal that had been the concern of European societies ever since the end of the Middle Ages but the absence of which was the very reason for the lack of change in traditional societies: reducing work. It could be argued that China did not direct its knowledge and techniques toward this end because of the vast labor force at its disposal. But to save labor is also to save time, and Max Weber has of course put great emphasis on the puritan attitude to time as a major factor in the rise of capitalism. Rational management of one's time, combined with concern to carry out one's professional duty, becomes "a necessity of the technical and economic conditions of mechanical and mechanized production"—time is money.[4]

In China, neither the traditional view of the world nor the social organization encouraged a different approach to time. Since the medieval period there had indeed been a form of state capitalism, but it was based on types of organization and behavior that ran counter to the innovations—as much social as technical—that combined to stimulate and be stimulated by the rise of industrial capitalism in Europe.

In analyzing these obstacles, the eminent sinologist Etienne Balazs suggests reasons that could just as easily be applied to many developing countries today—*mutatis mutandis,* naturally, because there has never been an exact equivalent of the Middle Empire in terms of its philosophy, its political organization, and above all its "celestial bureaucracy":

> What China most lacked in order for capitalism to grow was neither mechanical skill nor an aptitude for science, but the field available to private initiative. There was no freedom for individuals and no security for firms, no legal basis for the rights of others besides the State, no form of investment except in land, no protection against the arbitrary demands of officials or against the actions of the State. But perhaps the factor that paralysed things the most was the overwhelming prestige of the State bureaucracy, which from the outset hampered any efforts by the bourgeoisie to be different, to become aware of itself as a class, or to fight for an independent place in society.[5]

Over the long term in Europe, free-market capitalism has been the source of technical transformations geared to raising the productivity of labor—and also their raison d'être. These transformations were able to occur and to spread only because the economic and social context was favorable, and this context, itself altered by progress in science and technology, in turn influenced the pace and direction of technical innovation. The process was extremely complex, as David Landes stresses in his analysis of the consequences to be drawn from the history of the Industrial Revolution. The ingredients in the process were both economic and noneconomic, and the relationships among them were neither rigid nor imposed: "Each industrializing society develops its own combination of elements to fit its traditions, possibilities and circumstances. The fact that there is this play of structure, however, does not mean that there is no structure."[6]

This historical detour is useful if we are to understand the peculiar conditions of technological takeoff in the newly industrialized countries—it is indeed essential, and one may wonder why those who think about development have so often overlooked it. But the world as it is and societies as they develop do not fit the models of the economists or the visions of the prophets. "The time has come when models, by exploding, allow denials of all sorts. . . . When those who 'relied on themselves' are opening their gates to the multinationals. . . . When everything is in turmoil, the enemy is an abstraction, when curses lose their force and miracles vanish."[7]

This left-wing reaction is matched on the side of the free marketeers, with the same tone of surprise, in the Interamerican Development Bank's report on the evolution of economic theories in Latin America. It shows that none of the theories—such as structuralism, the theory of dependence, monetarism, export-oriented strategies—has taken full account of the diversity or the contradictions of the growth paths. "Reality has not been so clearcut. The basic relationships underlying economic behavior are not known with any certainty. Contradictory hypotheses have been around for a long time and they are rarely in harmony with the historical facts."[8]

Thus, after forty years of practical experience and theories, the failure of the purely economic paradigm, whether left- or right-wing, comes down to discovering that history exists, with its chances and necessities, and there are never either compete breaks or miracles but rather uncertain efforts to take off, always in suspense, based on a long period of coming to maturity. "The analyses and the models proposed for industrialization and development, liberal or radical, are weak because of their problems in adjusting to a reality which is characterized today by the importance of the historical dimension."[9]

Three Giants and Four Dragons

There are about thirty developing countries among the newly industrialized countries (NICs), depending on whether the list includes the Mediterranean countries (e.g., Spain, Portugal, Yugoslavia) and some of the eastern European nations. But far fewer are in a position to sell a range of manufactured products in international markets, especially goods involving medium or high technology—the figure then is barely 10 percent of the Third World.

In 1986, when OECD dealt for the first time with newly industrialized countries, statistics on the export performances of the NICs were available for only six of them: Brazil, Mexico, South Korea, Hong Kong, Singapore, and Taiwan. These figures show that between 1965 and 1973, their exports of manufactured goods grew on average by 30 percent per year, ranging from 50 percent for Korea to just over 20 percent for Hong Kong; between 1973 and 1982, this growth slowed slightly, to 20.7 percent. (In the first period, nine of the most highly industrialized OECD countries recorded an average growth rate of 16 percent; the rate was 12.7 percent in the second.) The NICs' exports did not go only to the OECD countries, but they also created their own export markets.[10]

The types of products exported are even more significant. Traditional goods (textiles, leather, clothing, and shoes) grew substantially but irregularly, whereas medium-technology products such as vehicles, chemicals, nonferrous metals, and high-technology goods (telecommunications equipment, sound recorders, electrical equipment) grew strongly and steadily. In 1964, such products accounted for a tiny proportion of OECD imports; by 1984, they had become well established, especially imports by the United States, Canada, Japan, and Australia. High-technology exports to North America tended to come from Latin America rather than Asia, whereas the opposite was true of medium-technology goods.

The NICs play an increasingly important role as suppliers of manufactured goods for OECD countries. Between 1964 and 1984, the proportion of low-technology goods imported from the six NICs covered by the statistics fell from 84 to 54 percent, medium-technology goods settled at about 20 percent, while high-technology goods rose from 2 to 26 percent. This does not mean, however, that OECD countries stopped exporting manufactured products to the NICs; on the contrary, such exports grew, faster (15.5 percent) than the rate of growth of exports within the OECD area. Nevertheless, whereas the share of Latin American countries was about the same as Asia's in 1964, by 1984 the Asian market had become three times larger than that of Brazil and Mexico together.

Among the NICs, three stand out for their geographical size, population, and resources: Brazil, China, and India. They are the only ones capable of mastering all aspects of the contemporary technical system. Those that are nicknamed "the four dragons"—South Korea, Hong Kong, Singapore, and Taiwan—have had great success in producing and exporting some products based on the new technologies, which puts them well ahead of Argentina or Mexico and even further in advance of the latecomers, Indonesia and Malaysia.

What do the three giants and the four dragons have in common? In contrast to the African countries, they all share a culture involving writing and printing. They all have a scientific past at the interface of traditional methods and European science and have long-established higher-education institutions modeled on those of the industrialized countries. And, unlike most former colonies, they have all been industrialized for a hundred years.

How far this is true varies from one country to another but above all from one section of the population to another within each country: Singapore and South Korea, for example, have among the highest proportions of engineers in the world. Yet whatever the differences, this inheritance has been the basis for building a pool of skills and talents over several generations (manual workers, technicians, engineers, scientists, entrepreneurs) trained to master the modern technical system. For both the little dragons and the giants, the dynamism that has grown out of this learning process cannot be understood unless it is seen in its long-term context.

The three giants differ among themselves not simply in their scientific and industrial ambitions but in their political organization, economic approach, and social choices. Capitalist Brazil contrasts with socialist India and communist China. In Brazil, everyone speaks the same language, almost everyone believes in the same religion, and, after a period of military government, the country is striving for a democratic system, so difficult to achieve in Latin America. India is a subcontinent, with a population divided by language, religion, class, and taboos, where parliamentary democracy manages to hold on. Despite communism, China remains what it has always been: a conglomeration of peoples and states the central authority has always had difficulty in controlling, where the tensions between conservative and modernizing Marxists seem to repeat, in the shadow of a new bureaucracy, the age-old hesitation of the empire between the need and the fear of opening up to the West.

Nothing links these three countries—not the political system, economic organization, or scientific institutions. Their historical relationships with the West have been very different. India has had constant contact, from Alexander the Great via the caravan trade under the Moguls to British colonization, with a cultural and religious side that has always held aloof from the influence of modernity. China has had extremely intermittent contact, from Marco Polo to the Jesuits, with Western "barbarians," who have always been treated with suspicion and whose entry in large numbers, to wage the wars against opium that led to the collapse of the emperors' power and the administration of the mandarins, will always be seen as an assault on traditional ways. In Brazil, there was systematic pillage by the Portuguese motherland until the empire was founded in the nineteenth century; then came the evolution of a nation that always modeled itself on the West, even though its Indian memory and African heritage set it apart.

In contrast with the Brazilian universities, influenced first by the Portuguese, then the French, and now the U.S. model, Indian universities have been patterned on Oxford and Cambridge; the Chinese institutions were reestablished after World War II along Soviet lines. But each of these countries has borrowed from the West the same notion of science and technology as an instrument of change; each has provided itself some time ago with government agencies, close to the leadership, to define, plan, and implement a policy for this area. From Brazilian positivism to the Chinese command model by way of the Indian mixture of socialism and neoliberalism, each in its own way practices a form of state capitalism and is interventionist in its approach to research and innovation, with the most active support coming from the military. This interventionism is responsible for much of the success achieved in science and technology by policies of modernization.

The example of Brazil is revealing in this regard. The first engineering schools and scientific institutions were created after the king of Portugal moved to Brazil in the early nineteenth century, after fleeing from Napoleon's invasion of the Iberian peninsula. Until that date, as in most European colonies, all industry was forbidden, from textiles to saltmining. From the middle and even more the end of the nineteenth century, from empire to republic, the state provided guarantees for capital investments and intervened directly in creating manufacturing industry in the geographic triangle of Rio de Janeiro, Belo Horizonte, and São Paulo. The frenetic upswings associated with gold, sugar, coffee, and rubber that shifted the focus of growth from region to region made clear how vulnerable colonial exports were. "In the name of progress they demand protection for manufacturing; in the name of nationalism, they spread the principle of industrialization."[11]

The depression of the 1930s and World War II generated a true internal market in Brazil, supplied by industries producing import substitutes. The capital built up made it possible to diversify productive capacity; new immigration from Europe and Asia increased the numbers of technicians and entrepreneurs; and transnational firms opened branches. According to Celso Furtado, southern Brazil became the industrial metropolis of a sort of British empire, with the Amazon playing the part of Africa and the northeast that of India. The transition period not only prepared the ground; above all, it prepared minds, with a synergy of intermingling cultures, awakening nationalism, and positivism directly inspired by Auguste Comte. As Stefan Zweig wrote in 1941, looking at Rio just before he committed suicide, "Brazil has learned to think in measures of the future"—and to think big.[12]

At the end of the 1940s, the creation of the Technical Center for Aeronautics was the start of a policy to foster new technologies that led, twenty years later, to the expansion of independent arms and computer industries. The center spawned the Institute of Aeronautical Technology, established at São José dos Campos, where generations of engineers and managers have been trained on the model of MIT (many graduates were indeed sent on to the United States, to Berkeley or Cambridge, to study for their doctorates). The aims were to move away from the imitative behavior found among engineers in most developing countries, to build an infrastructure of national industries, and to increase the qualifications and skills of the technical staff.

Nationalism combined with a concern to reduce the level of dependency on the industrialized countries, especially the United States, brought the military and the business community to uphold the same strategic view: The state, as banker, entrepreneur, and customer, should take the initiative in major programs for technological development, on the understanding that later, the industries thus created would gradually become independent of state supervision or at least would encourage private firms to act as suppliers (these latter were managed and often owned by graduates of the institute, the "Iteans").

The ground was cleared, plowed, and sown over a period of two decades; nothing was lacking—the center had continuous support, laboratory equipment, and courses in management from the best foreign schools to inculcate in the Iteans not so much an esprit de corps such as is found among graduates of the French Grandes Ecoles as "a relatively homogeneous cast of mind, based on the values of development and the nation, values which stress the country's technological potential and favor technical efficiency and competence."[13]

The harvest began to be reaped in the 1960s, when the government decided to reequip the armed forces by replacing cut-price materiel from the United States with weapons systems produced locally. The state investments encouraged the entrepreneurs, who from the outset aimed for products that could have both civil and military uses. EMBRAER, with its 300 subcontractors, and the firms of ENGE, AEROTEC, MOTORTEC, and others work within the framework of horizontal integration, which allows them not only to minimize the costs and the risks but also to develop an international network for marketing and maintenance.

Planning for aircraft and military vehicles is based on using materials, engines, and electronic systems exploited for civilian purposes by foreign firms, and each product is aimed at a particular export niche. For instance, Bandeirantes planes are meant for transport and Ipanema planes for agricultural uses. At the same time, the government sets up barriers to prevent importation of products similar to those being developed by local industry and operates a fiscal policy that gives generous tax breaks to all entrepreneurs who buy shares in public companies like EMBRAER.

The combination of state interventionism, public purchasing, and protectionist measures creates a new field for entrepreneurship, which removes the obstacles hindering technology policies in most developing countries. As Renato Dagnino has emphasized, basic research in the universities does not spill over into other technological spinoffs because of lack of resources for long-term projects and, above all, because "the agencies that provide support are more concerned to see their financial procedures being followed than to see the research they support achieve concrete results." Even when the universities carry out applied research, at best this leads to a nonviable prototype; and when they set out to achieve a technological breakthrough, the result does not interest local industry, which would rather buy products more cheaply abroad.[14] There is no alternative but to create a sector "outside the market," with firms tied to state patronage and projects whose costs at the outset cannot compete with those of foreign rivals.

Nonetheless, in the field of computers, university researchers linked up with Iteans, industrialists, and the military from the beginning; their skills in designing machines and especially in software were indispensable. Yet although the state policy has been identical, with the same measures aimed at isolation and protection of the domestic market, it is far from certain that it has had the same success. The problem with manufacturing is always that of ensuring that mastery of the production of new technologies occurs within national frontiers. And although it is possible to be closely connected with these technologies from the outset, as happened in the case of the sophisticated mechanical engineering required in the aircraft industry, such openings into the latest developments in electronics are increasingly limited.

Initially, the alliance between the technical and military experts involved a demonstration of "economic nationalism"—the concern was to bar the way against imported computers or those made locally by multinationals. From the beginning of the 1980s, there were enough local firms to encourage balanced joint ventures with foreign firms. The prices of Brazilian microcomputers have since fallen steadily, after starting well above those of U.S. competitors. In 1982 the first clones of Apple II cost 50 percent more than the imported originals; by 1984 the difference was 37 percent, and by 1986 the prices were the same, thanks to economies of scale and experience. However, there is still a significant gap as regards software, and the progress achieved by Macintosh is making it hard to catch up because the new integrated circuits have elements that cannot be copied.

The traditional policies of interventionism and protectionism are now becoming counterproductive. Unless there are massive state investments in order to help local firms to continue to master all the stages of production, there will be no alternative but to sign more license agreements or establish other links with foreign firms in order to stay in the race.[15] The promised land of technological autonomy is disappearing into the distance, as the speed and complexity of technical progress create obstacles for economic nationalism that are ever harder to overcome.

The contradictions inherent in the reserved market, which threatened to cut off the Brazilian computer industry from the latest technical advances, gave the experts pause for thought. "Who needs Macintosh?" asks Simon Schwartzman, arguing that the latest models meet the needs of the most industrialized countries, not those of developing countries.[16] Yet that model is a critical step toward a mass market, because the machines are user-friendly and accessible to nonprofessionals. The law protecting the Brazilian computer industry was passed in 1984; it has now been revoked and it is easier to import from abroad. It remains true, however, that state interventionism has prevented total domination by foreign firms; the capital now available in terms of skills and infrastructure gives the Brazilian computer industry more room for maneuver than most developing countries can boast.

The Cost of Interventionism

The most striking aspects of the Brazilian experience, which are also found in all the other NICs, are the state's willingness to intervene and the time taken for a project to mature. These two factors are inextricably linked, and both are necessary to enter and stay in the technological marathon. The examples of India, Taiwan, and South Korea are no more "miraculous" than that of Brazil. In addition to a long-established industrial base and an educated labor force, it takes time—several generations—and unwavering public commitment for the policy to produce results. Of all the NICs, South Korea and Taiwan have benefited most from foreign financing, especially the U.S. aid given free of charge. Yet there, too, takeoff occurred only because of an existing preindustrial base dating back to the nineteenth century; then between 1910 and 1945, during the Japanese occupation, heavy industry grew up around certain activities (mining, mechanical engineering, fertilizers) geared to the needs of the colonial power.[17]

There is in fact no shortcut to creating a relatively independent manufacturing industry, and if the state chooses to make this a priority with a view to the longer term, the decision is a political one, overriding all other economic or social considerations. The benefits of being a latecomer, able to adopt at once the most efficient technologies, are cumulative only if all other development aims are subordinated to raising skill levels and strengthening productive capacity in the target sectors. Moreover, the country must know how to exploit these benefits: Barely a dozen developing countries have entered the technological marathon, thanks to their energy and sacrifices, and the role of social discipline in their success should not be underestimated.

As Joan Robinson has pointed out, it is remarkable that the most rapid growth in the Third World has occurred under extremely repressive regimes.[18] The truth is that whereas basic research goes hand in hand with democratic demands—involving publication, discussion, and criticism—applied research and, above all, technological development tend to adapt well to "closed" political systems, in contrast to the "open society" that Michael Polanyi and Karl Popper argue is indispensable to the progress of science proper.[19]

At a deeper level, it should be remembered that basic research has had a very marginal role in all the forced marches toward closing the technology gap, from the Meiji Restoration to the Japanese buildup of the army between the wars, from military rule in Brazil to the tough regimes in Korea, Singapore, or Taiwan. Besides, democracy does not foster decisionmaking or planning by a small group of technocrats or senior army officers, who choose to target one sector for development at the expense of all the rest. Joan Robinson notes that there are also repressive regimes untouched by industrial growth, but an interventionist modernization policy based on the most advanced technologies does not allow opposition or, even less, anarchy.

Among the NICs, China is more remarkable for its potential than for its actual achievements. In the future, especially in the longer term, it could become the most dynamic of them all, provided the country does not succumb again to upheavals or turn in on itself. China, with India, has the longest history of scientific tradition and civilization. China is also the developing country with the most original development model as well as the most violent one. The peasant revolution and the seizure of power by Mao Zedong turned into an agrarian reform that rescued the peasants from the landowners and into a decision to shift and maintain the distribution of income in favor of the countryside at the expense of the towns in order to combat mass poverty. The Maoist slogan "Walking on both legs" meant keeping investment to the minimum wherever it was possible to use labor-intensive methods instead.

This is doubtless why so many developing countries still find the Chinese model so attractive: The most daring land reforms, everywhere except China, have always been subverted in the end by the large landowners (e.g., in Latin America), and state control of the economy has rarely diminished mass poverty. Nonetheless, the great march shows signs of a limp since the "four modernizations" were introduced by Deng Xiaoping in 1975, and decentralization, the freeing of price controls, and the concern for greater economic efficiency have created the greatest disparities in income since 1949 (except for the privileges of the *nomenklatura*). Indeed, Deng Xiaoping's reforms have been aimed directly at leftist egalitarianism.

In agriculture, the reforms have made it possible for smallholdings and rural industries to expand in circumstances that are closer to a free market than to Soviet-style planning. The people's communes and collective farms have been abolished, and the compulsory supply contracts reduced. Agricultural output rose between 1980 and 1985 by 67 percent (an average annual rate of 10.8 percent, compared with 3.5 percent between 1953 and 1980), benefiting above all peasant families living near large cities, who became major consumers of goods that had hitherto been denounced as luxuries (televisions, refrigerators, washing machines). In the meantime, 100 million peasants in the rural areas still live below the poverty line, and agrarian reform has not fundamentally altered the pattern of development in the countryside, which is based on labor-intensive methods.

The three other modernizations (industry, defense, science and technology) are occurring well away from the rural masses, in the coastal regions and the fourteen cities open to foreign trade. This is where Deng's reforms challenge the revolutionary model most strikingly, to the point that one wonders whether they will last. The opening up of this narrow strip of the country, like the concessions made by the last empress, is intended to lead to the rapid acquisition of the latest technologies and equipment, through direct imports, foreign investments, or joint ventures with foreign companies. The new system of "responsibility" for those in charge of production is supposed to make up for the low level of profitability and productivity of the enterprises, replacing revolutionary zeal with managerial drive.

The Chinese reformers of the end of the nineteenth century thought they could adopt Western methods without breaking with the system of traditional values—they argued that the body of Chinese principles should be preserved and simply applied so as to achieve Western efficiency. In the same way, the post-Mao reforms were introduced with the afterthought that revolution could (or should) be compatible with liberalization. In the beginning, the outstanding results of these reforms silenced the critics, after the years of autarky, scarcity, and chaos, that spanned the Great Leap Forward to the Cultural Revolution. Things soon began to go off course, however; prices rose and corruption surfaced. Deng's initial reaction was, "When you open the window to let in some fresh air, you can't stop the germs coming in too."

The risks of epidemic were nevertheless increased by the combination of three contradictory forces. First, there is the vast mass of the population, which had barely recovered from the horrors of the Cultural Revolution when it witnessed rising prices and the spectacle of some officials becoming involved in illegal businesses (the head of the Chinese-Japanese joint venture, Fujian-Hitachi, which makes color televisions in Fuzhou, was secretary of the local party and the trade union).[20] Second, there are the conservatives, the heirs of Mao and the Long March, who hinder the reforms because they threaten egalitarianism, or bureaucrats who fear losing their little patch of power. Last, there are the dissident intellectuals, writers, artists, academics, students, and scientists, who think that liberalization should bring a fifth modernization, democracy.

Wei Jing-Sheng, who made this suggestion, was at once arrested and sentenced to fifteen years in a labor camp, but the debate went on until 1987, when Deng Xiaoping had to back off: The policy of openness was put in abeyance, the secretary-general of the party was replaced, the president and vice-president of the Academy of Sciences were forced to resign, and many officials lost their posts in the campaign against "bourgeois liberalism." Later, the few signs of a new openness led to the repression of the Tiananmen Square demonstrators in 1989; once again liberalization was brought to a halt. Will China—closed to foreign influences after the failure of the missions started by Ricci in the seventeenth century, forced to open a crack by Western aggression in the nineteenth, turned in on itself again after 1949—enjoy even a decade of openness at the end of this twentieth century before returning to the autarky of the Ming and Qing periods?

One of the present authors, invited to Beijing in April 1986 by the State Council for Science and Technology, in his official discussions was constantly asked the same two questions: What "recipes" account for the success of the capitalists in the field of the new technologies? And if such recipes exist, how can the People's Republic implement them without losing its identity? It was not easy to answer that there was no quick fix, and that a country's capacity to innovate was usually linked to its form of social, political, and economic organization. At the time, the talk was all of decentralization, giving greater autonomy to both provinces and enterprises, and injecting a certain amount (but how much?) of competition and responsibility into structures that had hitherto been entirely state-run. But how far is too far? During that same week, the *China Daily,* the English-language official newspaper, reported that an "illicit dealer" in Japanese cars and televisions had been shot, though it was not made clear wherein his activities were deemed illegal.

Expanding upon Weber's interpretation of the rise of capitalism in Europe, Albert Hirschman has well argued that the Protestant ethic was not enough on the part of individuals; there had also to be a "paradigm of interest" on the part of the state—a shift from the violent and heroic behavior of feudalism and aristocratic regimes to the "quiet passion of interest" associated with middle-class merchants, bankers, and captains of industry. The paradigm of interest presupposes a world in which entrepreneurs derive the double benefit of predictability and stability and in which the state then makes limited interventions. It is not just a matter of achieving personal salvation through thrift, in the Calvinist manner, but of restraining the prince's whims in order to foster the city's expansion.[21] Now, as in the imperial past, Chinese entrepreneurs who are successful do not know whether they will be congratulated for making money or shot; they thus are understandably inclined to hesitate before taking any initiative or declaring their profits.

Deng Xiaoping has always insisted that whatever is borrowed from the West will not challenge the communist system, in particular the collective ownership of the means of production. But the first reforms did introduce a few experiments in private capitalist enterprise—admittedly very isolated but dynamic. What happens to Chinese communism if they influence the rest of the economy? And if they do not succeed, what is the point of the violent efforts to make up the country's huge technology gap? The 1984 patent law, the first of its kind in China, illustrates the contradictions: It was meant to encourage researchers and technical experts to be inventive, but article 6 stipulates that the rights belong to the unit where the inventors work, and article 14 further restricts whatever monopoly the inventors might enjoy by stressing that "the competent agencies of the state, provincial government, the autonomous regions and the municipalities have the power to designate which institutions are to exploit the inventions developed by any unit under their administration."

As one left the Stalinist building that houses the State Council for Science and Technology, one could not but marvel at these efforts by the ruling team to adapt Mao's peasant revolution to the social demands of capitalist innovation. China has realized that it lacks the capital, the technology, the technical experts, and the capacity to train new elites quickly, as well as the management experience it needs so desperately if it is to modernize itself. The statistics cannot hide the position: Out of the country's 6 million scientists and technologists, more than half have barely the equivalent of secondary-level education; only 6 percent are engaged in truly scientific research, and nobody knows how many of those are working exclusively on military research.[22]

Within the country as a whole, with its population of over 1 billion, there must now be proportionately more mathematicians, scientists, and technical experts in the Western sense than there were when Matteo Ricci was teaching algebra and geometry at the court of K'ang-hi. But the gap is enormous and all the more obvious when China compares itself with its closest neighbors—the former Soviet republics to the north, India to the southwest—and when it sees how far the four little dragons have gone along the road to modernization. These last have not, it is true, had to cope with the same problems of geographical size and population, nor have they had to deal with the aftermath of upheavals such as communist China has forced upon its inhabitants.

In 1986, during the "Beijing Spring," people talked openly of the price to be paid for modernization. In reporting a debate among social scientists, the *China Daily* noted that what was at issue was "no longer only to update science and technology. Social scientists today are appealing to the nation to modernize its system of values and mode of thinking." How to escape from the age-old influence of Confucianism and the more recent one of

Maoism without abandoning the essentials? "Modernization is seen as synonymous with Westernization, and Western culture was considered almost synonymous with decadence and decline.... The study of Western culture is to make for better assimilation, not exclusion"—so that openness means first recognizing the positive aspects of that culture.[23] Paradoxically, in the Chinese year of the rabbit (which followed the year of the tiger), the claws came out repeatedly, and the conservatives stood out against modernization.

And yet despite the repression in Tiananmen Square in June 1989, the policy of economic openness in the coastal regions continues. The communist model collapsed in Eastern Europe but survives in China as much thanks to the precommunist imperial heritage of the country as to the influence of Stalinism. After Deng Xiaoping dies, it is possible that the system will become more liberal and will ultimately implode in turn, like the Soviet system. But there is no guarantee that learning about the principles of the market economy will lead China to be more receptive than in the past to Western (i.e., foreign, if not "barbarian") influence, with its principles and values that start with respect for the individual.

In China, the social cost of interventionism aimed at closing the technology gap has been not only to challenge the communist identity of the country, just as the arrival of the Christian missionaries in the seventeenth century threatened the Confucian view of the empire, but also to abandon the development model that had previously been in operation (except in the strategic sectors of defense research) based on egalitarianism. Matteo Ricci came to Peking to sow the seeds of the Christian religion along with those of European science. The Chinese of those days, as now, realized that it was less a matter of converting to a faith than of assimilating alien social ideas. As a young philosopher said in the debate reported by the *China Daily* "For the purposes of reform, there is much in foreign culture that we can use. For instance, the sense of competition, creativity, independence and respect for individual ability and opinion."

10 History's Revenge

"Science and technology for development" is a slogan that has inspired both the programs of international organizations and the modernization policies of all the developing countries for a quarter century. The achievements of the newly industrialized countries cannot hide the fact that the slogan is a hotchpotch of prejudices, and that overall there have been more failures and examples of waste than successes.

The prejudices are paralleled by deceitfulness as well. It is not true that basic research is essential for development; that the most advanced technologies are geared to the needs of most developing countries; that the Information Revolution is a shortcut to making the whole of the economy more productive; that making industrialization a top priority is the best way to meet the challenges of development. The truth is that there is no shortcut: Science and technology help (perhaps) to speed up the process of modernization in a particular sector isolated from the rest of society, but they do not absolve any country from having to deal in the long term with all the prerequisites that must be satisfied in order to escape from underdevelopment.

Unequal Growth

For most economists, the growth of per capita GNP is an indicator not only of economic success but also of "welfare." What does success mean, without considering the inequalities that come automatically with the growth in GNP in most developing countries? According to this indicator, the situation of many developing countries has improved in the last twenty years. Yet the gap between rich and poor in most countries has remained as great as ever and in many instances has widened. This was the conclusion of Louis Emmerij, head of OECD's Development Center, in a 1986 report in which he stressed that the successes were the exceptions and that catching up can happen only in the long term—indeed, the very long term:

> While the OECD countries, including the least well off, continued to see their per capita incomes increase, Latin America and Africa regressed by a decade or more. In the case of these two continents, not only did income differences with OECD countries widen, but people and countries actually became worse off in absolute terms. Only Asia can hope to catch up with the OECD average—and this only with time. Indeed, if per capita income growth rates in both Asia and OECD remain at their rates of the past 15 years, Asia will not reach the OECD per capita income average until more than *two centuries* from now.[1]

It is thought that in choosing to industrialize, the country in general will benefit from the improvement of one sector, even if the others are neglected or left untouched. This approach reduces development policy to a straightforward technical job the sole aim of which is to increase the economic efficiency of the sector concerned. In the end, the social tensions and upheavals do not really matter: Whatever the risks of revolt or decline, is not economic growth the way forward?

Among the newly industrialized countries, Brazil has been one of the most dynamic, and whatever the difficulties ahead, its progress seems likely to be consolidated, provided that the economy is not crushed by the debt burden. As in China and India, Brazil's efforts to industrialize have been based essentially on trying to become a major power—not in the sense of the industrialized countries in general but as a member of the nuclear club. In all these cases, nationalism for strategic reasons has been responsible for industrial expansion and for substantial technological advances. The research and innovation policy has been aimed at the whole range of high-tech strategic areas, from nuclear energy to missiles and from aircraft to telecommunications. In security terms, Brazil's neighbors seem rather less of a threat than India's or China's, yet the Brazilian arms industry has been the most successful in world markets.

At the same time, the high priority given to manufacturing has created huge distortions in the civil sector, at the expense of agriculture and public services (education, health, transport). The deliberately unequal growth—between sectors and regions—has made for industries whose competitiveness may be envied by many developing countries and even some industrialized countries. Part of the economy has certainly taken off, and the grip of technological *dependencia* has been loosened. Nevertheless, the choice has ultimately made the whole economy more vulnerable and, as a consequence, has threatened the social consensus. GNP has continued to grow, but negative aspects abound: The country has been flooded with consumer goods that only a quarter of the population can afford; most peasants remain under the feudal control of the large landowners; mass poverty has not decreased; famine continues to afflict the northeast; more than 4 million street children live on nothing in the big cities; and the debt is so enormous that there is no way it can ever be repaid.

There is obviously no such thing as balanced growth for a developing country wishing to take off economically. Albert Hirschman was right to stress, more than a quarter century ago, that such growth is a mirage, but also that the pressures and tensions often connected with the growth process can be beneficial. At the time, he attacked above all the idea, spread by the mythology of the Russian Revolution, that large-scale industrialization—carefully planned and carried out on several fronts—is the surest means to takeoff.[2] From this angle, Iran under the shah or Algeria under Houari Boumédienne were in different degrees very vivid illustrations of the deadend to which such strategies can lead.

On the other hand, the counterexamples Hirschman used as a basis for his criticism are not much more convincing. Have the countries like Colombia (which he had studied closely) that carried out their industrialization in phases rather than all at once really skipped *many* steps? They prove the importance of the "linkage effects" to which Hirschman gave his name, but this alternative growth strategy—which meant putting the cart before the horse in certain chosen sectors or firms—also had its limits. At any rate, impatience to catch up cannot be pushed too far: Putting the cart before the horse creates new distortions, and the stimulus to economic growth enjoyed by certain sectors merely puts others at an even greater disadvantage socially.

All in all, the inevitable inequalities of economic growth generate pressures that can act as stimulants in the short term, yet can become so unbearable in the long term that they ultimately bring the whole process to a stop. Very few countries, so far, have been able to make these inequalities work in favor of the development of *the whole of society*—that is, without going too fast or too far in exploiting profitable opportunities. What is the elasticity of tolerance of growing income inequalities?[3] The question is worded in economists' jargon; in simpler political language it might be rephrased, At what point does pursuit of growth cause social unrest or even civil war?

As Hirschman himself recognizes, this is where economists are shortsighted: Free-marketeers in Brazil, monetarists in Chile, Marxists in China, or leftists in Algeria—they are all obsessed with the growth of GNP and overlook the social consequences in the long term.[4] In the end, as in Molière's play, the invalid is cured, but he is dead. It is not surprising if this approach invites caricature or is antidemocratic, and that dissidents must be silenced if not removed for pointing out that development may have other costs.

Impatience to catch up, and the mirage of Western science as the magic formula required to do so, caused people to think that the process of modernization mainly involved finding technical solutions. The scientism—or the lack of awareness—that inspires this approach ignores the time factor and assumes that the elasticity of tolerance for the spurts of change is infinite. The success of the strategies of industrialization by forced marches is similar to cavalry charges that gain ground; the approach puts off the real battle of development.

History always gets its revenge on economics: A development policy that neglects the problem of income distribution inevitably lays itself open to social upheavals later on and consequently to the repercussions when the authorities are forced to become tougher and more arbitrary. To reconcile the ambitions for economic growth with greater distributive justice is undoubtedly the surest way not only to avoid social upheavals but also to guarantee continuing growth. None of the technical solutions offered by science and technology can act as a substitute for this political and social imperative.

The Hard and the Soft

It has become a cliché to label natural sciences "hard" and social sciences "soft." Although the latter claim to use the methods of the former, they do not have the same explanatory power. For one thing, investigations can never be confined to what is strictly measurable and quantifiable. There is no standard technique that can be applied to the study of a specific situation; each one requires a special and fresh approach. It is impossible to carry out the types of experiment on individuals or groups that would be needed to refute a theory in the Popperian terms of the natural sciences. Psychological and social phenomena are rather more tractable than the physical world: Situations alter, because human beings and societies can act upon them as they become aware of them (often thanks to the social sciences), and they cannot be reproduced artificially.

At a deeper level, knowledge of the physical laws of the natural sciences does not affect these laws. This is the meaning usually given to Einstein's remark that God is subtle but not malicious: God does not alter these laws just because human beings try to discover them. This is what prevents the coexistence of conflicting theories beyond Popperian refutation and also stops scientific theories from becoming ideologies. The intrinsic feature of scientific laws does not apply to the social sciences; the behavior of the social "material" is altered by knowledge of the laws relating to it, which is why social theories can turn into ideologies.

This does not mean that the social sciences have no possibility of universal application. Because they deal with human beings and societies, the disciplines they cover can never be separated from the historical and cultural context in which they operate. Moreover, applying the rigorous methods and aims of science to individuals or groups will never yield answers to all the questions raised or solutions to all the problems, relying on a single theory that, in this area, will always prove inadequate. The social sciences have interpretations, if not distinctive and competing schools, because the analysis of human or social phenomena inevitably involves several—sometimes conflicting—conclusions.

The story of development economics is, as Albert Hirschman has demonstrated, one of the best examples of the difference between the natural and social sciences. In the natural sciences, it is perfectly acceptable to formulate a new paradigm, and it is the task of traditional science to check, apply, and extend the paradigm. In the social sciences, a new paradigm gives rise to research efforts of a similar kind, but it is almost at once subject to nuances, criticisms, and even demolition, which challenge the notion of the steady accumulation of knowledge that is characteristic of the natural sciences. The history of development economics since 1950 shows that there can be advances in the social sciences, "but on condition that intellectual progress is defined as the gradual loss of certainty, as the slow mapping out of the extent of our ignorance, which was previously hidden by an initial certainty parading as a paradigm."[5]

The truism that social sciences are "soft" because they are not "hard" does not mean that they cannot contain the truth. Like all science, they attempt to increase knowledge and understanding. Measurement and prediction do not have the same role as in natural science, but the social sciences are not necessarily woolly just because they are soft. Yet it is true that if research in the natural sciences is ahead of the technical possibilities—technology itself is often ahead of the current possibilities—research in the social sciences usually lags behind the evolution of social mores. The nature of the field rules out the creation of a "social technology," but the disciplines of the social sciences nevertheless provide, as do any other sciences, ways of knowing and means of action.

It is in relation to technology, however, that the hard and soft overlap to the point of being indistinguishable. If one concentrates on technology in terms of the materials and physical elements of which it is made, one fails to see that it also has intangible components. It is indeed a product of natural science and engineering but also of the social sciences and of society. The very evolution of the contemporary technical system, as the information technologies in particular demonstrate, reveals how inseparable the physical exterior of the machines is from the invisible interior—programs, organization, standards. The feature that distinguishes technology from hard science is that it interacts with human beings (this occurs often, though not always, via information); above all, the intangible aspect of information, however hard it may be as a physical quantity, is as much linked to human thought as to matter in that it is a *language*. In this sense, technology is *a social process like any other* and is not simply a combination of matter, energy, and information designed to carry out certain tasks that could be transferred from one part of the planet to another without "translation."

Technology transfer is first and foremost a transfer of *culture*. There can be no transfer without assimilation of the know-how required for mastery of use of the new system of machines, and this is even more essential for mastery of production. From this viewpoint, hard and soft are indistinguishable: As a consumer good, a piece of technology without an instruction manual is a museum exhibit; as an input to production, without an understanding of the calculations used in its design, it remains as impenetrable as a foreign language; as the product of a culture, it will be sloughed off like a graft that has failed unless there is a grasp of the features that make it possible, if not to reproduce the item, at least to make it work in the alien setting.

The laws on which Western science is based are universal, but the capacity to make these laws operational is not. For that to happen, there must be institutions, methods, and a society that share the same rationality. The technology that develops out of Western science—it is indeed now closely dependent on scientific theory—can be transferred if the receiving agents are able, as they "unwrap" the package of imported physical objects, to decipher and use all that is covered by the wrapping: not just the hard components (materials, frame, and other parts) but the soft ones contained in the scientific code, the work organization, and the value system.

In this sense, there is no difference between transferring a soft item like law and another that seems impeccably hard because of its physical appearance. In the 1960s, there was much talk of "development law," meaning a range of written legal rules that the newly independent countries were to introduce in the place of their customary laws. What was at stake in these discussions was in fact precisely the same issue as in the more recent debate about adapting technologies to meet the specific needs of developing countries.

Modern laws were imposed from the outside upon largely rural populations accustomed to unwritten legal codes. Where the new laws diverged too far from social reality, they were simply not implemented, leading to so-called fantasy laws, just as advanced technologies cannot be used by more than a tiny fraction of the population, with no effect on the development of the rest of society. Michel Crozier's observation that "you can't change society by decree" is true of France and even more relevant in Third World countries. The written laws modifying the status of women, family organization, or landownership have never been sufficient on their own to change behavior. Furthermore, the "presumption of knowledge of the law," just like that of imported technologies, is a total fiction.[6]

There are no halfway houses, especially because modern laws are formulated in a language different from that of the majority of the population, a language that in any case uses technical terms that are not easy to understand. Few lawyers are skilled in explaining these matters to the general public, and they tend to concentrate on the modern sector of the economy. The available intermediaries (tribes, clans, families) are more likely to resist the new laws. Modern law, like imported technology, is far ahead of social reality. There is no magic virtue in its very existence any more than there is in the availability of technology, which will bring about the social and cultural changes required for it to spread.

Development law is an option within legal policy in the same way as technology is an option within industrial policy, in the context of efforts to create radical changes in society. Besides, these transformations, however timid, will always take time to work through and be absorbed. Law is relatively static, or at least it changes gradually. By contrast, technology is dynamic and needs an increasingly large number of trained intermediaries in order to spread.

The introduction of written law, like that of imported technology, cannot be separated from its economic, political, and social context and, above all, from the level of education of the society in general. But whereas a country (other than a democracy) can manage without many legal experts, no country can function without a sufficient number of middle managers, supervisors, technicians, and workers able to make the most of the technical system on which it relies for survival.

Walking on Two Legs

The Information Revolution alters the position of the industrialized countries, but it does not fundamentally change that of the developing countries by giving them greater guarantees of being able to catch up. This is all the more true insofar as most developing countries are less concerned with catching up than with mastering the techniques required to produce the goods and services they need most urgently. In this sense, Mao Zedong's phrase about "walking on two legs"—relying on both traditional labor-intensive methods and modern capitalist technologies—will remain valid for the developing countries for a long time yet.

This slogan has meaning only as long as development does not start to limp too badly. The Maoist strategies of the Great Leap Forward and the Cultural Revolution adversely affected the modernization of traditional sectors while maintaining the priority given to building up industry based on the manufacture of the most advanced weapons systems. There are certain similarities with the post-Revolution policies followed in France under the Convention in 1792.[7]

Admittedly it is not science as such that can play a decisive role in the evolution of most developing countries: Well-trained technicians and middle managers are more valuable than scientists with doctorates conferred according to the standards of the international scientific community. The Republic needs scholars, but underdeveloped countries need technicians even more. This is absolutely essential to ensure widespread basic technical skills, based on mass primary education. Where agriculture is and promises to remain the preponderant activity, it is not possible to talk about science policy other than as a charade. Nonetheless, this is what did occur in many developing countries in Africa and Latin America with the assistance of the former colonial powers and international organizations. The result was to create in those countries large bureaucracies administering the semblance of research being conducted in a vacuum.

Science can help to make traditional techniques more efficient (for example, in biomass conversion or fermentation in general) by providing a better understanding of the fundamental processes involved, and it has a role in combating tropical diseases. But science of this kind must be geared to local conditions, or it is merely a prestige activity. Where it is less a matter of takeoff than survival, the model of scientific institutions and of science policy inspired by the most advanced industrialized countries will lead only to waste of human and financial resources. The diffusion of scientific methods in agriculture, food, and health requires first extension services and workers—not researchers.

In this regard, the time taken to spread a technical culture able to improve the productivity of traditional methods is the only area where a real shortcut is possible. There is no alternative to education, methods, and research that aim primarily to diffuse superior know-how throughout the population. For most developing countries, the burden of ignorance linked to poverty will not be lightened in expensive universities or in laboratories with ambitious projects that set out to rival the Western model. What will help is action on the ground, at the level of the actual problems of daily life, with the basic scientific and technical knowledge geared to the realities of the local situation.

Improvements in hygiene, housing, food, health, and employment have less to do with technology transfer, and less still with science per se, than with the endogenous capacity to increase the numbers of people with the training needed to cope in the local setting. This does not mean, however, that this training does not require instruction and even apprenticeship of a scientific type.

The fashion for "appropriate," "alternative," "intermediate," or "soft" technology gives the impression that the (often valid) criticisms of the vast technological programs signify a call for a return to Gandhi's spinning wheel. In fact, the idea is rather to upgrade traditional techniques by combining them with modern know-how.

Experience has proved, yet again, that recourse to even the simplest methods makes social demands. In developing countries where the labor force usually has no industrial background, a technology—however much simplified—remains a new technology. The end users must be involved in each aspect of the implementation of an appropriate technology, and no matter how apparently self-evident is the final solution adopted, it must be recognized that for the technology to be absorbed and to become economic, the users may have to master complex principles and techniques.[8]

The "alternative" movement started up in the 1960s in the West, in the aftermath of questioning about the consumer society, and was reflected in student revolts, the counterculture, challenges regarding economic growth, and increased awareness of the ecological constraints. It found in E. F. Schumacher, the author of *Small Is Beautiful*, a prophet influenced by both Buddhism and Gandhi who sought to encourage "not mass production but production for the masses." Schumacher set up the Intermediate Technology Group in London and sparked imitators in several developing countries (India, Pakistan, the Philippines), whose contributions to agriculture, building materials, tools, and equipment have been far from negligible.[9] Whereas Schumacher's followers in the West (e.g., environmentalists and opponents of nuclear energy) tend to preach the rejection of technology, in many developing countries they represent the "technologies of necessity."

In fact, India had already shown the way by developing large numbers of small plants geared to the needs of the local market. For example, there are three types of sugar factories: One type produces coarse cane sugar (*gur*), using small traditional mills powered by a couple of oxen (a technique introduced by Alexander the Great); another produces a more refined sugar that keeps better (*khandsari*), in small plants that process between 100 and 300 tons of sugar per year; and finally, there is industrial sugar production in large modern refineries, which supply only a third of total Indian consumption, indicating the significance of the traditional methods.

Although the methods are archaic, the *gur* matches precisely the needs and the productive capacity of a large part of the rural population. The *khandsari* plants are constantly improving their processing techniques in order to satisfy the demands of the urban population, and all the machinery is made locally by Indian firms. They use roughly twice as much labor as the modern refineries, but the latter must import some of their machines and constitute isolated pockets without linkages to the local economy.[10]

The small *khandsari* plants, like the thousands of Indian minifactories making bricks, cement, woodpulp, and paper, would have to compete with large modern factories if India could afford to follow the one-sided Western model aimed only at high productivity. Instead, they form a kind of complementary production system, which has the advantage of being better suited to the rural economy than are the modern plants. Without some protection through the tax system on both purchases of raw cane and sales of the finished product, they would be unable because of low productivity to withstand the competition from big factories. But the savings on imported machinery and above all the additional jobs provided cannot be underestimated in the cost-benefit analysis. Necessity rules—in the sense that social considerations oblige the Indians not simply to walk on both legs but to use as well the arms (or crutches) provided by methods that to us seem obsolete.

In any case, as Jean Gimpel has demonstrated, even the most archaic techniques can be improved to achieve greater efficiency and productivity by using lessons inspired by a technical system that the West has neglected for centuries. Gimpel started from a simple idea that in the end was given United Nations support: Most of the people in the 2 million villages of the Third World do not have the technical knowledge (particularly of mechanics) that European villagers of the twelfth and thirteenth centuries had. They are thus not one industrial revolution behind but two, because they do not know about machines driven by vertical wheels (mechanical tilt hammers, bellows, woodsaws, camshafts) or methods of lifting water (Archimedes' screw, wells with counterweights) that were invented or updated in the "first Industrial Revolution" of the European Middle Ages.

Gimpel had scale models of medieval machines made in modern materials (transparent plastic and chromium-plated steel) in order to emphasize that these ancient techniques did not belong to a past era. By explaining how his models worked to villagers in Nepal, Togo, and Kenya, he was able to show them how to improve their local methods. This technology transfer was carried out using direct instruction, in the tradition of the notebooks of Villard de Honnecourt, a thirteenth-century architect and engineer, and the results were all the more remarkable in that Gimpel's pupils learned to maintain and repair all the elements of the new technical system they adopted.[11]

"Who would believe that by going around with a model under your arm, you could help people to be better fed?" asked Régine Pernoud, who had the idea for the book in which this experiment is described. Perhaps all one needs to do is to observe, as Jean Gimpel did, that in this age of computers, rockets, and nuclear power, millions of households still use stoves without chimneys. The chimney was invented in Europe in the twelfth century: The modernized stoves not only use wood more economically, but they also improve living conditions, especially for women.

The Three Strategies

Despite the diversity of situations in the Third Worlds, three general strategies can be identified, albeit there are as many nuances and variations as there are countries.

At one extreme is the classic approach that aims at growth in national income through increasing investment, industrialization, and participation in world markets. In these countries, the process of industrialization rests on a Western-style higher-education system, the training and employment of a highly trained and specialized workforce, a complex program of scientific research, and technological development programs linked to military ambitions. A policy for science and technology that is in some degree aligned on the priorities of the industrialized countries may help to back up the spread of knowledge and the application of techniques as well as to ensure that the human infrastructure is there to deal with the transfer of the most advanced technologies that could not otherwise be assimilated under local conditions.

At the other extreme, development is seen in terms of covering basic needs rather than in terms of takeoff. For such countries, where agriculture will remain the dominant activity and plans for industrialization are necessarily limited, scientific research will be marginal, because the technical infrastructure, the financial resources, and the trained workers are lacking. A science policy in these circumstances would be an illusion. The prerequisite for any kind of development is to update traditional methods, by initiatives to increase and spread basic technical skills in line with local needs.

Between these two extremes are many mixed strategies that aim simultaneously to strengthen and develop applied research and also to support the expansion of primary activities through vocational training based on local needs. This means both drawing on the dynamism of traditional techniques and adapting some of the latest technologies to meet the most urgent needs of the local community. For these countries, the most pressing problem is not so much to train Western-style researchers as to increase the numbers of good middle-level managers through better primary and vocational schools with courses geared to the local situation.

As there are intermediate technologies, so there are intermediate levels of knowledge between the traditional and the scientific. The diversity of circumstances must be matched by a diversity of strategies. Nonetheless, development is put in question wherever the choice of technology means that only a minority benefits. Distortions of this kind arise as soon as industrialization is given top priority. The combined strategies not only reflect the problem of making a choice, but they meet the need to balance industry and agriculture. Agriculture requires growing support from a range of industries (fertilizers, cement, irrigation pumps, tools, and machines), but some industries also depend on agriculture (e.g., textiles, food processing). In short, any strategy for research and innovation must take into consideration the multiplier effects on all sectors.

For countries that have an industrial infrastructure, the problem is not to choose between heavy industry and small firms by sacrificing one to the other but to ensure that each interacts with the other, thus creating linkages beneficial to both. For example, small firms might act as subcontractors for the large ones or make finished products. Experience shows that it is fatal to switch from one extreme to the other: The drawbacks to prioritizing heavy industry and plans based on vast technological supercomplexes do not mean that soft technologies are necessarily the panacea. In fact, updating traditional techniques is probably the best way to strengthen the economy and create jobs.

The problem of making scientific and technical choices is related to the economic and social objectives each country sets itself, taking account of the special constraints it must face. Even more than in the industrialized countries, the programs of scientific research and technical education cannot be separated from the type of development path the country hopes to follow. In other words, what a country wants to do with science and technology is ultimately linked to what it cares about. R&D is merely a tool used toward goals other than science in the strict sense—those are the objectives that shape the future of a society, not the other way around.

Technical change and technology itself make up a social process in which individuals and groups make choices that determine the allocation of extremely scarce resources. There is a saying "Tell me who you know, and I'll tell you who you are." In the context of development and the use of scientific and technical resources, this might be rephrased "Tell me what you are seeking, and I'll tell you what you really care about." If the concern is with the development of the whole of society, then education must be the priority. There must be training not just of scientists and senior managers but training of workers in intermediate technical skills in all types of activity, with pride of place given to agriculture and craft activities rather than to manufacturing and services.

In this sense, although the Information Revolution provides new opportunities, it also threatens to lead the Third Worlds even further away from their true circumstances. A society in which services predominate does not match the immediate needs of the developing countries, and the speed of computers does not absolve them from dealing with the long term. The solution in any case cannot be found in the algorithms or the machine systems of Western rationality. Rather it is to be found in common sense or popular wisdom—it takes time. The best course is the one suggested by a Chinese proverb: "Give a man a fish, and he eats once. Teach him to fish, and he can feed himself for the rest of his life."

Mote and Beam

There is a final point we cannot avoid after this race "through the looking glass": By what right can we, citizens of the industrialized countries, criticize the developing countries' delusions of technological shortcuts?

There are several good reasons. The first is that the "well-being" of the industrialized countries did not materialize with a single stroke of a magic wand but rather occurred over several centuries and at the price of upheavals—social, political, economic, and cultural—that are all too easy to forget.

The second is that the countries now "affluent" are partly responsible for the several Third Worlds' choice of industrialization. We did not export just the model of the Industrial Revolution, with its hope of rapid takeoff; we gave them the vision of the consumer society, though it is far from certain that it matches their real needs. We should admit that in helping to spread this vision, the rich countries did not make themselves any poorer.

The third reason is that while the industrialized countries may appear "advanced" and certainly have dealt with basic needs, they themselves are now in serious difficulties. Among these problems, the persistence of high levels of unemployment and the large and growing pockets of poverty prompt reflection on the limits of the process of modernization, even after several centuries. In both industrialized and developing countries, the expansion of the new technical system based on intangibles—from information technologies to biotechnologies and new materials—is not in itself the means of salvation. Science and technology, defined by the political Left and Right as the miracle cure-all, are myths for us as well.

The survival and even increase in illiteracy in societies that have vast resources allocated to education, retraining, and technical training reveal that the threat of polarization is not confined to the developing countries. The mote should not make us lose sight of the beam: Despite relative prosperity, fairer income distribution, social security, education, R&D budgets, the expansion of brainwork, and the spread of computerization, the "creative society" that some see already replacing the "productive society" everywhere is in fact still a Promised Land.

The truth is that our societies also have their Third Worlds, and they are likely to grow. The scribes of the industrialized countries are already replacing their battered old typewriters with word processors. They will always have customers.

APPENDIX

The following three articles were originally published in *Social Science Information* as contributions to a debate organized by the journal in response to the French edition of this book. They are reprinted here with permission from Sage Publications Ltd.; the authors would also like to thank the journal's editor, Mrs. Elina Almasy, very warmly for having launched the debate and for her interest and assistance throughout.

All references to page numbers have been altered to correspond to those of this book. However, readers should note that the wording of the quotations from the text given by Christian Comeliau and Amilcar Herrera are those of the original articles and may differ slightly from the present translation.

J.-J. S.
A. L.

Symposium on Science, Technology and Development: What Options for Third World Countries?

The subtitle we proposed in our call for contributions was: "What options for Third World countries... Perpetual dependence or the advantage of being a late-comer?" We opened our symposium in the June 1989 issue with the review article by Christian Comeliau on the challenging book by Jean-Jacques Salomon and André Lebeau, *L'écrivain public et l'ordinateur—Mirages du développement* (Paris: Hachette, 1988).

In the review article the discussion was already launched, since counter-arguments by Salomon and Lebeau in response to critical comments by Comeliau were included. It was followed by the first invited contribution we received—by Michael J. Moravcsik (USA). In the September 1989 issue, we carried a paper by Marcel Roche (Venezuela), in December two papers by Amilcar O. Herrera (Brazil) and Zhao Fu San (China), in June 1990 the contribution of Amiya Kumar Bagchi (India) and then, in September 1990, a paper by Jacques Gaillard (France).

Finally, in the December 1990 issue, Jean-Jacques Salomon and André Lebeau commented on the contributions and drew the conclusions of the Symposium.

The editors of Social Science Information

Editorial slightly adapted from *Social Science Information*, Vol. 29, No. 4 (1990), p. 840.

Is the Third World Headed for Perpetual Dependency?

Christian Comeliau
Commissariat Général au Plan, Paris

It is not the primary purpose of scientific research to respond to social demand, and when a developing country is successful in establishing a programme of high-quality fundamental research, the effect on its state of development is zero; such research is therefore not a priority for development strategies. This intentionally provocative thesis is the centrepiece of a stimulating book by two authors who are exceptionally experienced in the area: the first founded and for twenty years headed the Science and Technology Policy Division at the OECD; the second, who is now director of the Service de Météorologie Nationale, was the assistant director of the Centre National d'Etudes Spatiales and then of the European Space Agency. Which means that their thesis deserves, at the very least, serious examination and discussion.[1]

In these notes, I limit myself to outlining the discussion, taking up some of the ideas which form the skeleton of the thesis and indicating the questions (for the most part mentioned by the authors themselves) they raise for those in charge of scientific and technological policy in Third World countries.

The first idea is that there exists something called a "universal" science. This science developed more specifically in the Western World, but it claims to transcend culture: "the universality of Western science derives from its postulation of laws of the universe which are constant in time and space" (p. 54).

The primary purpose of this type of science is not to respond to "social demand" (p. 69), and even less to the objective of development; but it can be thought of, together with technology, as "the instrument of change" (p. 133). This becomes understandable if a distinction is made—and here we have the second essential idea—between wanting to "know" and wanting

Reprinted with permission from *Social Science Information*, (Sage, London, Newbury Park, and New Delhi), Vol. 28, No. 2 (1989), pp. 438–444.

to "do," which leads to the traditional distinction between "fundamental research" and "applied research," even though it is growing increasingly difficult to draw a hard line between the two (p. 55).

At this stage in their reasoning, the authors seem tacitly to accept a point they do not go into, but which is postulated by what follows: fundamental research exercises no direct control over how its technology is used; or, to put it more exactly, if such a thing as universal science does exist, the research which produces it does not, in principle, place any restrictions on the scope of its possible technological applications. In which case, one could speak of scientific research as having a plurality of technological potentialities.

But a third point must be kept in mind: a new distinction between "controlling the production" of and "controlling the use" of technology (p. 26). The "new technical system" makes these two types of control much more complex and therefore more costly to acquire, especially control over production. A new form of inequality is therefore emerging which puts Third World countries at a disadvantage, first in terms of their capacity for scientific and technical research, but also in terms of jobs: from this point of view, industrialized countries must rush to keep pace with technical progress while developing countries suffer from loss of jobs with no new jobs to compensate (pp. 103–104).

But what produces this inequality in the capacity to deal with the cost of scientific and technological progress and their consequences? One obvious cause is financial; but others are structural: the unfolding of this progress and even more its effects on development are tightly linked to the structures of the societies which adopt them—social and cultural structures, systems of preferences and collective values. "The structures and organization of society define the grounds on which scientific knowledge and technical innovations contribute to growth—and not the other way around" (p. 5); the most binding tie runs therefore from social structures to the impact of scientific and technical development even if it is hard completely to deny a counter-effect of scientific and technical development on social structures, something which the authors assert later in the text: "Society and technical change take turns moulding each other" (p. 19). With the result that, while "universal" science may have developed historically in the West, all cultures are not equally ready for it (p. 65). It has even been remarked that, because of their cultures which are "outside the bounds of, if not blatantly foreign to, the intellectual foundations" of the new technical system, Third World countries are "sentenced to become passive victims of its effects" (p. 87). This means that scientific and technical development will be aimed first of all at fulfilling the needs of those societies doing the research, that is advanced industrial societies.

The preceding arguments inevitably lead to a series of questions or rather policy choices confronting those in charge of scientific and technological policy in Third World countries seeking a strategy of development. If fundamental science has no immediate social use, if technological pluralism is possible in theory but costly in practice, what types of scientific and technical research should be encouraged? Would it be better to abandon all attempts at producing scientific knowledge and stick to studying its technical uses and applications? What is the cost of the various forms of control over production and use of technology, not only in financial terms—which are in themselves prohibitive for a great many countries—but also in terms of jobs, and especially in terms of cultures (giving up at least certain aspects of value systems)? What "short-cuts" may be taken on the way to scientific and technical development, and what are the conditions?

Concerning the above questions, the thesis of the work remains highly pessimistic, as we have said. The authors consider that those in charge have neither the *means* to undertake fundamental research, nor have they any true *need,* since this research is done elsewhere, is published and available, and since, at any rate, this is not what provides solutions to the concrete problems of development that confront decision-makers. They should therefore stick to "applying the *available* scientific paradigm to creating technical capacities" by developing "a program of applied research"; "in the short and middle term [they] do not need to worry about expanding or enriching the scientific paradigm" (letter from André Lebeau to Christian Comeliau). But it is on these perspectives, or rather on the way they are presented, that I find the authors' reasoning not entirely satisfying.

My main objection—the problem which seems to have "vanished" in the course of the argument—is the following. Science does not necessarily respond to social demand, fine. The answer to this demand comes from technological development, filtered in a complicated way by the economic, social and cultural structures of the societies in question—whence the orientation of today's technological development. Of course one may well wonder whether it corresponds to the urgent needs of any society whatsoever; but the fact remains that those who decide obviously act on behalf of the "developed" countries, that is precisely those societies which control the production of science. Technology will be called to respond to the needs of these societies, and not to those of developing societies. We have here an example of specialized, contingent technological development situated in an historical context. For even if science is a universal phenomenon, the possibilities for technological development are dependent upon the *degree* of scientific advancement and thus on the *content* of acquired knowledge at a given moment in history. It would be difficult to argue that this content is universal.[2]

Under these circumstances, it is easy to see the disadvantages, except in the short run, for Third World countries of scientific and technological policies which renounce participation in the production of science. The authors' reasoning would be acceptable if it were possible to envisage a complete dissociation between fundamental and applied research which would make it possible to pick up any given piece of applied research from a common trunk of fundamental scientific knowledge and reapply it to totally different needs. But, as we have said, this dissociation is becoming increasingly difficult. Would it not be better then to accept the logical consequences of this negative state of affairs and consider that—even if the principle of the existence and legitimacy of a "universal science" is admitted, as it has been defined here—social and cultural structures determine not only the use to which science is put, but its very production, or rather the way its production is organized? For in practice this postulates choices, priorities, preferences which make it impossible that the concrete, precise content of scientific knowledge acquired at any given time in history—as distinct from technical knowledge for the same time—go uninfluenced by the society which produced this knowledge and therefore by the interests of that society.

If this be the case, developing countries cannot renounce *contributing* to the progress of this universal science; their participation will of course be modest at first (for all of the above reasons), but in the long run it will influence the content of science utilizable by all human groups. On the other hand, will not countries which renounce such a contribution on the unfortunately confirmed grounds that "science does not contribute directly to development" end up in a situation of perpetual dependency, with no hope of a brighter future?

Salomon and Lebeau accept this implicitly, speaking of increased levels of dependency and showing that countries such as Japan today feel the need for greater participation in theoretical scientific research. If they mean that gradual control of the production of science comes through control over its use, we can but agree; and it must be admitted that the cost of research excludes most developing countries. The principle of demanding immediate participation in this research stands, however; for renouncing scientific research will in all likelihood only aggravate the problem, in particular because renunciation deprives these countries of the training opportunities that come with research. Furthermore, the authors consider "education and general basic training policies" to be the true priority, emphasizing that the type of scientific research developing countries should "renounce" would at any rate be the "marginal" kind and, in support of their thesis, they invoke the example of Japan (letter from Jean-Jacques Salomon to Christian Comeliau).

As for discussing the practical modalities of a Third World strategy of participation in scientific research, this would make little sense in the context of the general remarks I have chosen to present here. Among other factors, these modalities obviously depend upon the size and the developmental level of the country concerned; India, China or Brazil can contemplate strategies which are in no way comparable with those of small African countries (see the typology suggested on pp. 29–30 and passim). But the authors stress that the "immense majority" of developing countries have no "research facilities, no scientific legacy, no industry capable of integrating the results of research, no qualified work force, etc." (letter from J.J.S to C. C.). One question dwelt upon by the authors deserves particular attention, however, and that is, is "leapfrogging" a possibility? Can a country without the means to make a frontal, general contribution to the production of science skip the intermediate stages and specialize directly in certain scientific work or in the production of certain kinds of advanced technology?

There are numerous examples of this, particularly in the three large countries mentioned above, in areas such as weapons production, space research, computer science, etc. By placing stress exclusively on the danger of getting the "cart before the horse" (p. 3), and on the global cultural dimension of "technology transfer" (for example, pp. 152, 154), might the authors not be underestimating the practical uses of such a selective strategy in getting a national collectivity started in international scientific research? The fact remains that such a strategy can be seen neither as an all-purpose "shortcut" nor as a means of "catching up" (p. 13); that "programs of scientific research and technical training are inseparable from the kind of development a country claims to be pursuing" (p. 158); and, above all, that "forced industrialization" carries a very high social pricetag—this is true anywhere; and that "the prestige enjoyed by the universities and science in certain developing countries is far from answering the most basic needs of the majority of the population in these same countries" (letter from J.J.S. to C. C.).

One last remark. Although it does not explicitly exclude them, this book does not deal with the social sciences, an omission which can easily be explained by the difference in the problems raised, or by the legitimate reticence one may feel in using the same word, "science," to designate two fundamentally different epistemological approaches. However, the need for research in the social sciences in and by developing countries themselves appears even more urgent than that for the hard sciences, though many countries are not willing to look into it. Certain passages in their text on "cultural pauperization" as linked to the cultural domination produced by computers, as well as the fact that "many developing countries buy back from abroad information concerning their own national reality: dispossessed of their history, they must pay to see it" (p. 118), lead me to believe the authors would agree. In an admittedly different area, this observation can only reinforce the need to go beyond techniques inspired by scientific production one does not control.

Notes

1. A preliminary draft of this review was submitted to both authors, who were kind enough to send me their reactions. I have tried to include their comments in the present version, sometimes quoting directly. I alone am, of course, responsible for any misunderstandings or errors which may remain. I should also like to thank Ignacy Sachs for his suggestions.

2. André Lebeau has indicated to me that he is in total disagreement with this proposition, which he considers to be erroneous. He writes: "The universality of Western science resides, as does its operative capacity, in its mode of production (p. 54); this means that, at a given moment, the science which is 'produced' is universal. Of course science is not a finished product, it progresses in scope and insight, but the universality and operative capacity of what is acquired remain intact."

Obviously I do not intend to resolve the question, but it seems to me that the discussion would do well to look at something that T. Kuhn wrote more than a quarter century ago concerning "scientific paradigms" and "scientific revolutions" (Thomas S. Kuhn, *The Structure of Scientific Revolutions,* The University of Chicago Press, 1962, second edition, 1970). He remarks that what goes into making up a "paradigm," for a time, contains its own limits: "A paradigm can (...) insulate the community from those socially important problems that are not reducible to the puzzle form, because they cannot be stated in terms of the conceptual and instrumental tools the paradigm supplies" (p. 37). He then goes on to define scientific revolutions as "those non-cumulative developmental episodes in which an older paradigm is replaced in whole or in part by an incompatible new one" (p. 92).

The Advantages to Being a Late-Comer to What?

Amilcar Herrera
Director, Instituto de Geociencias, Unicamp, Brazil

Recent change in perceptions of the Third World

The book which is the origin of this Symposium, *L'écrivain public et l'ordinateur—Mirages du développement*, by J.-J. Salomon and A. Lebeau, reflects clearly the perception of the *problématique* of the Third World held by a considerable proportion of the key decision-makers in the central countries. This perception is at the root of the profound change in the nature of North-South relations which began in the early 1970s.

During the first two decades of the postwar period, the world experienced a renewed sense of *solidarity* based on the concept of "modernization." This concept of development relied on the belief that poor countries could attain adequate living standards through the economic aid of rich countries and, above all, through the activation of their productive apparatus by the introduction of modern technology. Such a view of development was readily accepted by the ruling classes of the Third World countries, because it seemed to offer the possibility of attaining prosperity without the need to make major socioeconomic reforms.

At the beginning of the 1970s several new factors emerged which contributed to modify the nature of North-South relations. On the one hand the first signs of the world recession appeared, making central countries pay more attention to their internal problems at the expense of their concern for those of the Third World. On the other, there was the combined effect of the growing awareness of the failure of modernization strategies in social terms and of the existence of environmental limits to development at the global level.

This new perception of the world *problématique* is clearly reflected in some very influential works published in the early 1970s (particularly *The*

Reprinted with permission from *Social Science Information*, (Sage, London, Newbury Park, and New Delhi), Vol. 28, No. 4 (1989), pp. 823–840.

Limits to Growth by D. H. Meadows and others, and *The Population Bomb* by D. and A. Ehrlich). According to those authors, mankind was facing an imminent global ecological catastrophe due, above all, to the uncontrolled growth of the Third World population.

As a consequence, the Third World started to be regarded by central countries no longer as that poor segment of mankind which could attain reasonable living standards through joint participation with the advanced countries, but rather as a threat to the material progress of the latter. Prospective studies carried out later in developed countries *(Interfutures* by the OECD; *Presidential Report on the Year 2000* by the US government; the *Brandt Report)* seemed to confirm such perspectives, insofar as they forecast that at the beginning of the next century the relative situation of the majority of the Third World population would be the same or worse than it was in the early 1980s.

With such a frame of reference, Salomon and Lebeau try to identify the reasons for the failure of most of the Third World in solving the problems of underdevelopment, and using their wide knowledge of the field, they concentrate on the scientific and technological dimensions of change. They make a controversial but rich and interesting critical analysis of the conceptions and strategies of development prevalent in a great part of the Third World. We strongly disagree, however, with their characterization of the obstacles which hinder the adequate use of science and technology by the Third World countries, and with their policy recommendations for the future.

Given the prestige of the authors, and the fact already mentioned that their views of the problems of underdevelopment have considerable support in the central countries, their opinions deserve careful discussion in the Third World. In our treatment of the questions raised by the Symposium, we offer a critical analysis of their position on the issues we consider most relevant to the debate.

The central thesis of the book is that technological advance, however revolutionary, cannot induce per se the social transformations that are the prerequisite of development. The structure and organization of society, on the contrary, determine the conditions under which technological innovations can contribute to development. Scientific research and technological innovation can be effective only where social structures, institutions and mental habits have previously eliminated the majority of the "blocking factors" which are characteristic of a traditional society (p. 5).

Starting from those premises, the authors make an analysis of the factors that, in their opinion, hinder the creative incorporation of the new technologies in the countries of the Third World, and of the possibilities of overcoming or neutralizing those obstacles. The results of the analysis are very pessimistic. Although the authors do not pretend to make long-term forecasts it is obvious from their conclusions that, if the conceptual framework on which the analysis is based is correct, the dependence situation of the Third World will continue, and will most probably worsen, at least in the foreseeable future. The most appropriate scientific and technological strategy for developing countries would be, according to the authors, to abandon basic research—in other words the generation of new knowledge—and to concentrate on applied research based on the available scientific knowledge.

Cultural obstacles?

The obstacles to incorporation of the new technological paradigms by Third World countries can be divided broadly into two categories: cultural, and socioeconomic and political. The separation is important, because the second category includes social dimensions that can be clearly identified and evaluated, and which can be changed or modified through sociopolitical action. The vaguely defined cultural obstacles, on the other hand, are very difficult to identify properly or to evaluate and, more important, can only change through long-term, poorly understood processes.

In our analysis we consider first the cultural obstacles. Although, according to the authors, Western science is "universal," not all cultures are prepared to accept it—"it is clear that any culture whose conception of natural phenomena is not based on mathematical rationality, is not prepared to adopt the scientific approach" (p. 65). Due to those cultural differences most countries of the Third World would be in an unfavourable position in relation to the new technological system: "external, if not totally alien to the intellectual foundations of that system, they are condemned to receive passively the effects."

The above is based, in our opinion, on the lack of a clear differentiation between the various elements that constitute a culture.

As is well known, science was recognized as an area of knowledge that could be clearly differentiated from philosophy only in the nineteenth century, with Auguste Comte's formulation: the study of natural, mental and social phenomena by empirical methods. So defined, science is not a Western creation. The generation of knowledge based on empirical evidence has been essential for the survival and progress of mankind since the beginning of civilization. Modern science represents an enormous advance over the past, but not a discontinuity in the basic principles and attitudes that have guided the generation of scientific knowledge since the emergence of *Homo sapiens*. The intellectual foundations of modern science, therefore, are not external to the cultural traditions of Third World countries. The main cultural differences in the field of knowledge are religious or philosophical, and they refer essentially to the ultimate nature of the universe, and not to the "operational" knowledge required for an effective interaction between mankind and the physical world.

The case of Japan helps to clarify the point under discussion. In Chapter 1 of the book the authors, discussing the possibility of the Third World countries reaching the stage of development of the central countries without repeating their long trajectory since the beginning of the Industrial Revolution, question the common concept that Japan, despite being "yesterday" a feudal country with an essentially rural economy, is now one of the most advanced industrial countries. In their view, that "yesterday" is very relative; Japan, through the revolution of the Meiji restoration (1868), entered industrial civilization only a little later than France (pp. 13–14). The present situation of Japan is therefore the product of a long effort of more than a century.

The interesting point about this presentation of the Japanese experience is that it supports a position, in relation to the problems of *rattrapage* and of the cultural obstacles to development, which is the opposite of that defended by the authors. What is important in the Japanese trajectory is not that after entering the Industrial Revolution Japan was able to compete successfully with the other advanced countries, but the fact that in little more than a generation it evolved, in the second half of the nineteenth century, from a feudal rural society to a society which had fully incorporated the Industrial Revolution—a process which took the Western European countries several centuries to complete.

We do not pretend to generalize the whole of the Japanese experience, but the basic elements which made it possible are certainly of general validity. The first is that Japan's belonging to a culture supposedly external to the intellectual foundations of modern science did not prevent it achieving total assimilation of Western knowledge and technology in little more than a generation, as demonstrated by the results of the Russo-Japanese war (1904-5). The second but decisive element was the political will of the Japanese ruling classes to introduce the political and socioeconomic reforms required to transform Japan into an industrial country comparable with the Western powers. As the authors rightly remark, the key instrument which made that transformation possible was the educational policy implemented after the Meiji reform.

The Japanese case—others could be cited—clearly confirms that cultural differences are not obstacles to a society's being fully able to incorporate modern science and technology, when there is the political will to implement a social project which makes an explicit demand of science and technology. What a great part of the population of the Third World lacks is information, and this is the reason why educational policy was so central to the Japanese experience. Through an adequate educational policy, Japan provided its population with the necessary information to enable it to incorporate and manage modern knowledge and technology, while maintaining its cultural identity.

Sociopolitical factors in S&T development

Although, as we have tried to show, the cultural specificity of the Third World countries is not an obstacle to the creative incorporation of modern knowledge and technology, it is undoubtedly true that most developing countries have made little effort to create R&D systems adequate to their needs.

In order to define an effective strategy of development for the Third World, we need to answer two interlinked questions: why do most Third World countries have R&D systems so weak and so poorly articulated to their social demands? and to what extent can the resulting technological dependency explain the difficulties of Third World countries in the socially creative incorporation of modern technology?

A brief look at the past will help us to understand the present situation. The new wave of innovations is but the culmination of the process of technological change that began with the Industrial Revolution. Those innovations started to enter the Third World at the very beginning of the expansion of capitalism, but the so-called process of "modernization" acquired its real momentum only after the Second World War and the ensuing wave of decolonization. The technologies introduced during that process also held the promise—as do the present ones—of more and better-distributed wealth. In the central countries, the benefits of increased productivity reached the great majority of the population. In most of the Third World, however, the impact of the new technologies was very different. Benefits have reached only privileged minorities, and the majority of the population still live in conditions which are not much better than before the beginning of the process.

The causes leading to the frustration of hopes in the Third World are many and complex; we have selected for brief consideration only those most directly associated with the application of science and technology to development. We take the case of Latin America, not only because it is the region we know best, but also because it is a particularly illustrative example. Latin American countries have been politically independent for more than 150 years, and thus did not have the problems which in other Third World countries could be attributed to the historically recent direct control of the colonial powers.

The reasons for that failure have been amply discussed using different conceptual approaches, the most important being dependency theory, which attributes the main cause of persistent underdevelopment in the Third World countries to their mode of insertion into the international economic structure. The *direct* mechanism responsible for the failure was the incapacity of the countries of the region to introduce the institutional and social changes required for the full incorporation of the wave of innovations, and particularly the redistribution of income.

The redistribution of income in the developed countries, as a consequence of the introduction of the mass-production technological style, had two main effects: to enlarge internal markets enormously, and to change the pattern of product demand. The market that was previously divided between luxury and staple goods evolved to meet the demand of the middle-income sector, which included the middle class and a considerable proportion of the workers, and which comprised the majority of the population. In most of the Third World there was no significant redistribution of income; in most countries, particularly in Latin America, there was a continuous concentration of income in the upper classes. The most important consequence, from the point of view of the productive structure, was that the demand for non-staple goods came almost exclusively from those minorities with an income equivalent to the upper and middle classes of the advanced countries.

It is clear from this that the main cause of the social failure of the countries of the region in the incorporation of postwar innovations was not technological dependency, but rather the socioeconomic strategy adopted to incorporate them. The weakness of the local R&D systems was only a contributing factor, *and more a consequence than a cause of the failure.* The imitative style of development did not create a significant demand on the local R&D systems, and so *there was no stimulus for the implementation of an active, systematic policy for science.* When this development model reached its limits, and a demand for more local scientific and technological inputs started to appear, the neglected R&D systems were unable to give an adequate response to that demand.

Analysis of the results of the postwar process of modernization in Latin America confirms what we have already seen in nineteenth-century Japan: the main factors which determine how a society incorporates science and technology are not cultural—in the sense used by Salomon and Lebeau—but sociopolitical. The explicit technological demand of the model of society adopted by the ruling classes conditions the way in which scientific and technological knowledge is incorporated and used. The cultural context—value system, philosophy, religion, mental habits—is an essential determinant of the *basic objectives* of development, but when those objectives represent the aspirations of the social body and are adopted by the ruling social forces, these cultural elements have never been a serious obstacle for the adequate use of science and technology to attain them.

Late-comer to what?

We can go now to the questions posed by the symposium: *perpetual dependency? or the advantages of being a late-comer?*

To be able to answer those questions, we need first to answer a previous one: *the advantage of being a late-comer to what?* In other words, what are or should be the objectives and goals of the conception of development which is emerging from the present world crisis?

To Salomon and Lebeau the objectives seem to be reasonably clear, although they never state them explicitly: they are the objectives implicit in the model of development being followed by the industrialized countries. That model or strategy would be the most appropriate one to enter the options of human development offered by the new innovations. They criticize some shortcomings and dangers of that strategy, but they do not question any of its basic premises.

This conception of the long-term strategy for world development, together with the almost complete disregard for the political dimension as a dynamic agent of change, is the main weakness of the line of reasoning followed by the authors. It does not take into account the real character of the present world crisis, which is, in our view, the ultimate requisite for the viability of any long-term world model of development.

In order to be able to answer our question we briefly consider the character of the present world crisis, in order to discover the central characteristics a long-term world development strategy should have in order to be viable.

The present world crisis

The elements of the crisis are many and complex, but those that are central in determining its trajectory and outcome are a style or philosophy of development based on indiscriminate economic growth, and an accelerated process of global impoverishment that affects more than two-thirds of the population of the world. The combined effects of irrational consumption in the advanced countries and the ecological impact of poverty in the Third World are deeply affecting the renewal potential of natural resources and the planet's ecological basis for development.

From the point of view of the viability of the present social and international world order, the most important manifestation of global poverty is the continuously increasing international inequality. The magnitude of the gap that separates the developed countries from the so-called Third World has never been so great in modern history. At the beginning of Western expansion, with the ensuing process of colonization, the average living standard of the population of the colonized countries was not much lower than that of the European ones. Now, the difference, measured in terms of material consumption, is in the order of about 1:10 or 1:15. As important as its numerical value, or more than that, however, is the *qualitative* change being produced in the character of the gap. At the end of the Second World War, the development objectives of both central and peripheral countries were, to a certain extent, similar. In developed countries, particularly Europe, poverty was still a problem, and part of the population had not reached an acceptable living standard. Servicing those needs was therefore a common objective of developed and developing countries alike, although their starting points were different.

Today the situation has changed radically; for most Third World countries the satisfaction of the basic needs of their population, in other words the attainment of the benefits of the industrial society, is still their fundamental objective. The central countries, on the other hand, are entering what is being called the "post-industrial" society, a stage of development whose *problématique* is very different from the one developing countries still confront.

As a result of the change in the character of the gap, even the dialogue between the two blocs into which the world is divided is becoming more difficult every day. This happens at the same time as the degree of interdependence between nations at all levels—political, economic, cultural, biophysical—is higher than at any time in the past. Unless an environmental policy at the planetary level can be implemented—and this is impossible with the present division of the world—there is little chance of avoiding a global ecological collapse in the not-too-distant future.

The elements of the crisis mentioned above imply the possibility of conflicts. The form and extension of those conflicts are conditioned by the fact that we have a destructive nuclear capacity equivalent to more than one million Hiroshimas. Besides the danger of collective suicide, the cost of the arms race is one of the obstacles to the solution of the problems associated with poverty. In 1985 the global military expenditure, almost a trillion dollars, exceeded the total income of the poorest half of mankind.

Summing up, if present trends are maintained, at the beginning of the next century we will have a world in which an estimated 80 per cent of the population will be in the Third World in living conditions which will be—according to all the forecasts based on projections of present trends—more or less the same or worse than now. At the same time the environment, under the combined and mutually reinforcing effects of consumption and poverty we referred to earlier, will be in a continuously accelerated process of global deterioration.

It is obvious therefore that the tendentious scenario, the one implicit in the work under consideration, is not viable, whether we consider it from the sociopolitical or from the physical point of view. It can be said that Western civilization has become *dysfunctional,* in the sense that it is no longer able to give adequate responses to the problems generated by its own evolution.

The only way to avoid the dire consequences implicit in the present world trends, is to formulate alternative development strategies more in accordance with the aspirations of the majority of mankind, and with the possibilities and constraints posed by our present understanding of the physical universe in which we live.

How the future desirable society could be

In order to describe in its basic characteristics the possible future society which could be a positive outcome of the present world crisis, we need to have a general idea of what their main conditions could be. In our opinion, based on the ideas elaborated in the Technological Prospective for Latin America Project (TPLA),[1] they would be the following.

1. The problem of the environment and of natural resources will be one of the determining elements of the viable development options in the next decades. We are aware that we will have to put a limit to the growing pressure on natural resources and the environment in order to attain an ecologically sustainable society. We also know that those resources are enough for the indefinite subsistence of mankind *provided we accept an austere material life*. Austerity does not mean deprivation; it means the conscious limitation of our consumption of natural resources to a level compatible with their retative availability, and with the preservation of the global equilibrium of the biosphere.

2. In a society whose almost sole objective now, above all in the Western world, seems to be the indefinite growth of consumption, austerity might appear an unavoidable sacrifice hard to accept. However, austerity, and this is a most important point, has a positive aspect which far outweighs its supposedly negative side; together with the advance of technology it will allow the reduction of the human effort required to provide for the material needs of life, and thus an increase in the time devoted to more creative activities. It could lead finally to the almost complete elimination of the social division of labour, the evil that, together with material scarcity, has been directly or indirectly at the basis of all social conflicts from the beginning of history. The existence of an upper limit to material consumption implies that only an egalitarian society can be stable, because that restriction would clearly show the fallacy of the present illusion that inequality at the social and international levels can be corrected simply through indiscriminate economic growth.

3. Another essential condition of a viable society is participation, which means the right of all members to participate in the social decisions at all levels. Besides being a fundamental human right, participation is an essential prerequisite for the attainment of a society really compatible with the environment: a global ecological policy, which implies a change in our pattern of consumption and in our whole attitude towards the

environment, cannot be imposed from above; it can only be effective if consciously assumed by the whole social body.

On the basis of the above, we can now define what should be the basic characteristics of any society which could be a positive outcome of the present world crisis:

(a) essentially egalitarian in the access to goods and services;
(b) participative: all members have the right to participate in social decision-making at all levels;
(c) autonomous (not autarkic)—this means the capacity to make decisions based on its own aspirations and possibilities;
(d) intrinsically compatible with its physical environment.

In other words, the compatibility should be based not on *a posteriori* corrective measures, but on the very nature of the style of development. The preconditions for the attainment of that compatibility have been discussed already.

These characteristics may seem too general, but they are sufficient to define a basic type of society and, more importantly, they represent goals shared by the majority of mankind. They are what can be called first-order, long-term goals which constitute the frame of reference for the formulation of the short- and medium-term objectives. Those objectives could vary greatly depending on the national, cultural or regional conditions and on the selected strategy, but they should fulfil the requisite of contributing to—or at least not hindering—the final attainment of the first-order goals.

The fact that the proposed society is possible, desirable and necessary does not mean that its implementation would be easy. The alternative represented by the present trends could prevail, with the consequences already discussed. However, the growing awareness at world level that the society implicit in those trends is not viable suggests that that trajectory is not as unavoidable as it sometimes seems.

We do not believe that only the Third Word is questioning the viability and desirability of the present world order. In the previous analysis there are unavoidable simplifications, the most important one being the apparent clear-cut division between North and South perceptions of the world. The attitude towards the future that is prevalent in the upper levels of political decision-making in most developed countries is far from being universally accepted. There is an important sector of the population of those countries—particularly significant in the young generations and among scientists and intellectuals in general—that strongly questions the vision of the world implicit in those perceptions of the future. On a more general level, that contestatory attitude is one of the elements of the grassroots, ecological, peace and feminist movements and, as we also know, a most important part of the literature on alternative futures is being produced in the developed countries.

All in the same boat

Now that we know where we want to go, we can come back to the question about the possible advantages of being a late-comer. As could obviously be expected, there is no definite, unambiguous answer to the question.

From the point of view of the initial material or instrumental conditions, the advanced countries have the same advantages over the Third World countries as they have today: a much more developed economic and scientific and technological capacity. From a wider sociocultural perspective, however, the one that is essential in the determination of the central *objectives* of development, the relative situation of the two blocs is different.

For a majority of the population of the advanced countries, to accept an austere society, in the sense already defined, means to renounce a social objective which in the Western conception of progress has been central since the consolidation of the Industrial Revolution: the indefinite increase in material welfare, conceived basically in present times as material personal consumption. For most of the countries of the Third World, on the other hand, the meaning of the alternative society could be very different. First, because being "late-comers" to the present supposedly universal conception of development, the value system on which it is based to a great extent, has not been wholly incorporated by them. Second, because for most of the population of the Third World such "austerity" would mean a level of material well-being undreamed of in the present situation.

A dimension of the process of change, in which the relative positions of North and South are also different, is the task of building a world order capable of implementing a co-ordinated and effective environmental policy to reach stable global equilibrium with the biosphere. This objective requires a component of international solidarity which transcends the purely technological and economic spheres.

Due to their economic and technological superiority, the industrialized countries have obvious advantages over the Third World to lead the process of world unification. There are other elements, however, which could hinder that leadership, such as a strong tradition of nationalism, a feeling of superiority over other cultures given by a long and recent period of world domination, and the fact that an equitable solidary world society would mean the abandonment of part of the privileges that the central countries enjoy in the present world order.

In the case of Third World countries, conditions are different. The new world order would be an obvious benefit to them, and they do not have the sociopolitical and cultural obstacles we mentioned for the countries of the North.

A last element to be mentioned in relation to the "late-comers" is that some of the great cultures of the Third World have preserved rich cultural traditions that have a less reductionist view of the universe and of the man-nature relationship than that prevalent today in most of the Western world. This wider perception of reality could perhaps contribute to the construction of a new, more solidary, world order.

Summing up, it is clear that, in the construction of a society that could be a positive outcome of the present crisis, each of the blocs into which the world is divided has advantages and disadvantages, and that the scientific and technological dimension, although very important, is not the decisive one. Given the complex interaction of sociopolitical, cultural and technoeconomic elements, it is very difficult to evaluate the relative weight each of those elements will bring to the evolution of the process.

In our view, finally, the outcome of the process will basically depend on two interlinked factors:

1. The realization, above all by the Western world, that in the crisis of survival now confronted by mankind there are no separate solutions for the North or for the South. We are now in a situation in which only the mobilization of the world around some basic common objectives can modify the path that is leading us to self-destruction.

2. The recognition that the main obstacles preventing Third World countries benefiting from the social options offered by the new wave of innovation are not cultural but sociopolitical.

TPLA science and technology strategy

The TPLA project, as we have already mentioned, is centred on the technological and scientific dimensions of change. It intends to identify the main trends of technological change which will predominate in the next decades, and their social, economic and cultural impact on Latin America. The final objective is to contribute to the formulation of a science and technology strategy adequate for the future development of the countries of the region.

According to the methodology of the project, a prospective view of the impact of the new technologies should start from two basic premises. The first is that the impact of the new wave of innovations on society can be properly evaluated only in the context of the present world crisis or, better perhaps, process of transformation. The second premise, closely related to the first, is that the character of the social impact is not solely determined by the nature of the technologies per se, but also, and mainly, by *the socioeconomic strategy adopted to incorporate them.*

The central consequence of those premises is that the basis for the formulation of the R&D strategy should be *the scientific and technological demands of the proposed society.* This approach implies a sequence of steps: (a) definition of the character of the desirable society; (b) identification of the obstacles to the attainment of that society; (c) formulation of a socioeconomic and political strategy for the overcoming of those obstacles; (d) determination of the strategy's R&D demand.

The time horizon considered by the project—the next three or four decades—is the *period of transition.* The objective in this stage is to create the necessary preconditions for the access to the new society. The transition is gradual; it is possible to say that it has been completed when the values and the forms of organization and social action of the new society clearly predominate over those of the present one. Although the final objectives are the same, the trajectory to the new society will be different for the North and for the South, due to the differences in their initial situations.

The proposed R&D strategy for the region focuses on two interrelated objectives:

1. The creation of national or regional R&D systems of a level and flexibility comparable with those of the industrialized countries. The objective should be quality, rather than size: the R&D systems of the central countries vary greatly in size and overall resources. This is a long-term goal whose attainment, starting from the present conditions of most Third World countries, would take, assuming a lucid and sustained effort, between twenty and thirty years.

2. To satisfy, as much as possible, the R&D demand of the socioeconomic strategy of the transition period to the new society. This objective requires a careful determination of scientific and technological priorities, the reorientation of the effort as a function of those priorities, and the institutional restructuring of the R&D systems.

Those two objectives imply the need to make compatible the policy for science with the policy of science. As is well known, most Third World countries have had a policy of science. The failure of most of those policies is due not only to the structural problems already discussed, but also to the almost total absence of a clearly defined long-term policy for science.

We cannot cover the content of the policy of science, but we briefly discuss here the basic principles that constitute the frame of reference of the R&D policy. The most important one refers to the problem of *technological dependency.* To begin with, we have to accept the fact that, in the short and medium term, the region will have to import most of the technologies required for its development. Only in the long-term future—two to three decades—will the region be able to attain a degree of technological autonomy comparable with that of the advanced countries. For this reason it is an essential prerequisite for the formulation of any R&D strategy to have a clear idea of the implications of technological dependency in relation to the objectives of socioeconomic development.

Technological dependency is not an all-or-nothing situation: there are degrees of dependency. Practically all countries have to import technologies in certain areas or sectors. The difference between the developed and developing countries is not only in their degree of dependency but also, and more important, in the policies they apply to import technology. The developed countries, capitalist as well as socialist, adapt the imported technologies to their own aspirations and possibilities. On the other hand, most developing countries in the capitalist world introduce foreign technologies to create deformed copies of the societies where the technologies originated.

Although, as stated earlier, it will take two or three decades for the countries of the region to attain a technological autonomy comparable with that of the advanced countries, it should be stressed that they already have degrees of freedom and that—if appropriate policies are implemented—the capacity to make autonomous decisions in the scientific and technological field will increase continuously in the period of transition. To maximize the benefits accruing from that expanding capability, a central task of the strategy should be to determine R&D priorities based on the socioeconomic goals. In other words, the problem is not to close the "technological gap" in absolute terms, but to gradually reduce it as a function of the demand of the socioeconomic strategy.

A central problem, therefore, of an R&D strategy is how to select technological solutions—endogenously generated or imported—adapted to local aspirations and possibilities. A general frame of reference can be given by the concept of *technological space*. This concept—developed in a previous project—is simply the systematization of well-known but often forgotten principles. It starts by stressing that technological problems that are identified as obstacles to development can only be understood in their true dimensions by taking as a starting-point the socioeconomic context in which they are immersed. In other words, a technological problem is always a component of a much wider "problem situation". From that point of view the problems we are concerned with can be divided broadly into two categories: those in which the technical solution is not presently viable owing to socioeconomic or political constraints, and those for which a technical solution is viable.

A typical case belonging to the first category is the problem of undernutrition or hunger in many countries of the region. The technological dimension is not determinant: the necessary technologies are available, and the local R&D systems have enough capability to cope with at least the basic problems. The root of the problem is not technology but the distribution of income. Another problem of this type is the deterioration of some agricultural productive ecosystems. The technologies required for a rational exploitation of those ecosystems are known, but cannot be applied unless a radical agrarian reform is implemented.

With the information gathered in the previous process, a set of assumptions or paradigms will be derived as the frame of reference for the final step of developing the required technology. The set of assumptions—which will contain scientific, technological, environmental, economic, social, psychosocial and anthropological information—will define a *technological space* which is basically the set of requirements and constraints that the technology has to satisfy.

In finally building the technology, all possible solutions that fit the technological space should be considered. As is well known, from a certain body of scientific knowledge many technological solutions to a given problem can be devised. The existence of an adequate frame of reference allows the exploration of a multiplicity of possible paths, and the selection of the one best suited to the particular situation.

It should be emphasized, finally, that in this approach endogenous generation of technology refers to the process through which the characteristics that the technology should have are determined. The *endogenous is the process of definition* and not necessarily the technology itself, which can be imported, provided it is appropriate. In this way, *the transfer of technology becomes an integral part of the process of technology generation.*

The R&D strategy briefly summarized is based on the conception, already stated, that only the project of an autonomous society—which is a *political* decision—generates the explicit demand that can activate the scientific and technological system, leading finally to technological autonomy. In other words, technological autonomy is a central tool for autonomous sustained development, but at the same time is a result of that development.

Finally, a brief reflection on basic scientific research. In the proposed R&D strategy for the Third World countries basic research has the same role that it has in the developed countries. Without basic research there is no possibility of any significant technological autonomy, but our divergence with Salomon and Lebeau in this matter transcends the problem of the application of science to the attainment of socioeconomic goals, and refers to the very conception of development.

Third World countries should do basic research not only because it is necessary for economic progress, but also because science, as art, is a part of the human adventure and development means, ultimately, the full participation in that adventure.

Note

1. The TPLA project is sponsored by the United Nations University and the International Development Research Centre (IDRC) Canada. The work is co-ordinated by a Committee composed of G. Gallopin, R. Dagnino, H. Vessuri, A. Furtado, P. Gutman and L. Corona, under the chairmanship of the author who is the project co-ordinator.

Science, Technology and Development

Jean-Jacques Salomon and André Lebeau

The debate taking off from our book *L'écrivain public et l'ordinateur* (1988) which this journal had the happy idea of initiating has attracted the interest and sometimes the criticism of readers much concerned with the problems of the Third World, especially the ones connected with producing and using scientific knowledge in developing countries. We are very grateful to the journal's editors for having launched this debate, and we appreciate the comments generated by this exchange, which have been both courteous and valuable. Some of the comments nevertheless suggest that we have sometimes been misinterpreted and hence misunderstood. The subject is too serious and the consequences too important for the debate to be based on anything other than what we actually wrote in the book—it should not be about what we are thought to have said but did not, in fact, say. The book is to appear in English next year and it may be that the misunderstandings and wrong interpretations which arose for readers of the French will be avoided in the English edition.

The following questions were raised by the journal on the theme and the content of our book: "What options for Third World countries . . . Perpetual dependence or the advantage of being a latecomer?" These questions obviously touch on one of the key issues in the book, but the response must first take account of the fact that the developing countries cannot be treated as if they were all identical. Indeed their differences are even greater with regard to their scientific and technological potential, and the ways in which they use it. In spite of the efforts to appear united (for example at the United Nations conference on "Science and Technology for Development" in Vienna in 1977), there is not *one* Third World but several, and their circumstances are so different that generalizations should be avoided. In science and technology, Brazil or Korea have absolutely *nothing* in common with Chad or Colombia or Sri Lanka. For purposes of simplification we proposed a classification with five categories of developing countries based *primarily* on their resources in science and technology. We then tried, on the basis of these categories, to understand and explain why some countries are more successful than others in exploiting their scientific and technical know-how, and above all why, *despite* the expansion and spread of the "new technologies" (computers, biotechnologies, new materials), the gap between the industrialized countries and most developing countries (the vast majority, in fact) threatens to widen in future. Our book set out essentially to explode the illusions of "shortcuts" or ways of catching up more quickly which the new technologies tended to encourage. We also wanted to point out that development policies relying on "forced marches," emulating the model of the industrialized countries, may well help to modernize certain sectors, but they lead to distortions and their social costs are very high.

> The mirage of Western science as the magic formula for catching up caused people to think that the process of modernisation mainly involved finding technical solutions... History always gets its revenge on economics: a development policy that neglects the problem of income distribution inevitably lays itself open to social upheavals later on and consequently to the repercussions when the authorities are condemned to become tougher and more arbitrary (pp. 149–150).

Naturally no classification can satisfactorily take account of all the economic, political, social and cultural factors affecting the development process. The relative strength of the various factors cannot be *measured*, and *none* can be removed without risk of wrongly identifying the ultimate causes of the "lag" or the successful "take-off." We stress that "purely economic definitions of developing countries tend to be distorting mirrors" because we believe that many other factors must be considered, including such things as "the stability of the government, how authoritarian the policies are, the class structure and the occupational structure, landownership patterns, skill levels in rural and urban areas, how widespread technical knowledge is, the investment in education, how long the universities have been established compared with other countries, the quality of training in the public and private sectors, the deficiencies of the central administration, etc." We stressed that this is a far from exhaustive list of the ingredients which are not strictly economic and which, when combined in an elusive but indispensable mixture, determine a country's development chances (p. 31).

All these nuances are necessary in order to understand the rest of our analysis of the problem and the reasons behind our pessimistic conclusions: science and technology are certainly not the panacea for problems of development. More specifically, we stress the fact that the process of modernization is not simply a matter of finding technical solutions. We wrote—recalling that Gunnar Myrdal was of the same mind in his last writings—that "just as the explanation for underdevelopment cannot be reduced to quantifiable factors, the struggle against underdevelopment is not just a matter of economic remedies." Having rejected this "economist's paradigm," we asked ourselves the question "What have the newly industrialized countries (Brazil, India, South Korea or Taiwan) got in common, in spite of all the differences as regards history, culture, political organization, type of economic or social system?" Clearly, there is no single answer to this question, but certain conclusions can be drawn from examining these countries' experiences *in comparison* with those of other developing countries. Ultimately the conclusion is what common sense suggests: to use scientific and technological know-how—and *a fortiori* to generate it—means fulfilling various conditions which are met in varying degrees in some developing countries but are far from being satisfied (still in highly relative terms) in most of the others.

For example, all the "newly industrialized countries," unlike Black Africa, have a tradition of writing and printing. (In the eighteenth century, the rejection of printing for religious reasons undoubtedly contributed to the scientific and technological decline of the Ottoman Empire.) Secondly, they all have a scientific tradition mixing traditional methods and European science, and some have a scientific heritage that is far older than Western science. Thirdly, unlike most of the other former colonies, industrialization started almost a century ago, and they all have universities with a long tradition of scientific and technical exchanges with the best research institutions in the industrialized countries. These shared features highlight the importance of *time* and *continuity* in the creation and development of both academic and industrial institutions able to open the way to relative technological independence.

Of course, the strength of these traditions varies from country to country, as well as within each one and from one section of the population to another (it is characteristic of underdeveloped countries that social conditions are highly stratified). "Yet it is effectively on the basis of combining these traditions that each has constructed, in the course of several generations already, a pool of talents and skills—manual workers, technicians, engineers, scientific entrepreneurs—'trained' in mastery of the modern technical system" (p. 137). In other words, for the newly industrialized countries *now,* as *previously* for Japan and certain European countries (for instance Norway and Sweden), technological "take-off"—the impetus that is generated by learning to master the technical system available—cannot be understood unless it is seen against the *historical background.*

But the time factor by itself is not the whole explanation. In our view, two other features explain the results that the newly industrialized countries have achieved today and that Japan and the "laggard" European countries achieved in the nineteenth century. The fourth prerequisite is that all these countries conducted their modernization policy on the basis of some form of state capitalism, encouraging both investment in education and the construction of an industrial infrastructure, because the top decision-makers realized early on that scientific education and knowledge could play a very important role in the development process. Last, *but not least,* all these countries have shown a collective desire to liberate themselves not merely from the political dominance of the most "advanced" countries but also from their technological dominance.

The examination of these experiences shows the complex interdependence of the technical, economic, political, social and cultural factors involved in the process of "modernization." In rejecting any kind of determinism—whether technical or economic—we argued that social and technical change are both cause and effect in the development process: the ways in which societies adjust to these changes are no more systematic than technical change is static. We constantly emphasized that technology is not simply a matter of "hardware"—it also involves people, organizations and management methods; it is therefore a *social process* which shapes society as much as it is shaped by society.

Three themes can be clearly distinguished among the main reactions to the book. We discuss them here, partly to clear up any misunderstandings that perhaps arose from failure to see what we had really said, partly to put our conclusions more bluntly, and also to stress where we disagree. The first of these themes concerns basic research, the second the relative importance of the various factors which might explain why certain countries "lag behind," and the third examines the very concept of development.

Science and development

Is it really "provocative" on our part, as Christian Comeliau (1989) suggests, to *state* that basic research (i.e. purely academic science where results are not aimed at short-term use and whose applications often occur only in the long term) has no direct impact upon the economic development of the Third World? It is characteristic of the failure to understand our book that most of our critics (especially Christian Comeliau and Amilcar Herrera) use the words "science" and "technology" interchangeably, without seeing (or wishing to see) that our analysis is based on an interpretation of what science and technology are which makes such a confusion of the two terms impossible. Although technology long predates modern science—making and using tools is one of the things that distinguishes man from other animals—science as we now know it is relatively a *very* recent phenomenon, and it is *even more recently* that scientific theories and discoveries have given rise to industrial applications. Nowadays, the advances of the technical system are rooted in the progress of the scientific paradigm(s), so that one can indeed quite properly say that technology is henceforth inextricably linked to science. Yet, for one thing, this has been true for only a very short period (barely a century), and for another, this development is *also* related to the industrial environment of which it has been both cause and effect. What use was Newton's theory of celestial mechanics for a century and a half? Or again, to take a more modern example, there was a long interval between Alfred Kastler's discovery of the basic principles of lasers (optical pumping) and the first practical applications by Townes, and an even longer delay before lasers were used in industry. So that although the Japanese can currently produce first-class lasers, they made *no contribution whatever* to the theoretical work which was required for these applications.

In fact we are dealing here with a major semantic problem because these distinctions are neither trivial nor unbiased. If you take no notice of our interpretation of what science is—if you fail to distinguish science from technology, the search for knowledge from use of know-how—then you are not talking about the same issues. Our critics all confuse the things we keep separate (see especially pp. 49, 55): technology is peculiar to human beings, it defines *homo faber*, whereas science corresponds to a period, a culture and an approach that are not only later, they are also utterly *different*. This approach produces information (scientific publications) and not a direct impact on things or people, and the influence that it may have upon economic and social development involves a process of transformation, of "translation," which takes a long time and happens in complex and unpredictable ways. The calculations of Dirac or Einstein had *nothing* to do with technology, even if later on they found practical applications in industry. Contributions to the progress of a scientific paradigm must not be confused with the subsequent development and elaboration of applications and innovations resulting from the discovery or formulation of principles generated by this paradigm.

This theoretical work is characteristic of basic research as it is conducted, encouraged and spread in the best labs by researchers vying with one another in the "international scientific community." Very few developing countries in fact possess the essential infrastructure—in terms of human, physical and financial resources—that would allow them to contribute in anything but an extremely marginal fashion to "the advancement of science." Jacques Gaillard (1988) confirms this statement in his remarks about the researchers in developing countries and provides fuller evidence in his thesis (1989).[1]

Furthermore, we neither said nor even hinted (as Amilcar Herrera seems to reproach us for having done) that developing countries should give up basic research altogether. Where there are laboratories and research teams that are *competent* (i.e. recognized internationally) and that have proper backing (i.e. are acknowledged by state authorities), basic research is not only, as Herrera says, "like art . . . part of the human adventure," it also plays an important role in education and can contribute to scientific progress. But let's be serious: how many labs and research teams in the developing world really satisfy these two prerequisites?

What we are talking about here is science as it is understood and promoted by the "international scientific community"; its criteria, subjects and goals are, in fact, virtually all set by the laboratories—both public and private—of the most advanced industrialized countries. We must therefore stress to Christian Comeliau that the idea of a science which is "relevant" and geared to the "demands of society" is a contradiction: science of that kind would simply not fit the reality of the scientific establishment and the activities of researchers in the "advanced" countries. There is "good science," and the rest is of no interest to the "international scientific community." Let us go further: this community is rarely interested in subjects of research that are relevant to the special problems of developing countries. We wrote that science is by nature elitist: science in the poor countries, just as in the rich ones, does not try to help the poor, but instead sets out to advance knowledge and satisfy the curiosity of the scientists. The tropism of "international science"—the attraction it holds for researchers in developing countries, for intellectual or financial reasons—has two unfortunate consequences: too much research in Third World countries is conducted in areas that have nothing to do with their most pressing problems, and too few scientists take an interest in these problems, squandering intellectual effort on other things.

Surely it is obvious, as we said, that it is characteristic that the education systems of underdeveloped countries produce too many people who are overqualified in comparison with the available resources and infrastructures (in particular, the universities), and far too few technicians and middle managers in comparison with the real needs of society. The rapid expansion of institutions of higher education, which has happened without a matching growth of vocational training to improve the level of intermediate technical skills, is making this distortion far worse. One has only to think of India, which produces enormous numbers of scientists but which also has one of the highest rates of unemployment and emigration of scientists in the world. We came to the conclusion that the kind of scientific and technical training widely found in the developing countries based on the model of the best universities in the industrialized countries, is not best suited to local circumstances as regards its contents or its aims. Is this a provocative conclusion? It should be recalled that at its outset, the Industrial Revolution owed little to science and to "scholars," and the expansion of industrialization in Europe was based in the first place on the spread of primary education to the whole population rather than on a vast increase in the numbers of university students.

Basic research certainly contributes to developing the skills needed to understand the principles behind the most sophisticated technical solutions, but it is not essential simply to make them "work." It is truly indispensable for teaching science *in the making* to those who are making it, to make known the most recent advances to those who are themselves in the vanguard of scientific discovery, but it is not absolutely necessary in order to teach scientific principles—*ready-made* science—to those who will use them in practical ways in their jobs. Surely knowing how to use this type of knowledge, which nowadays naturally requires a background in science, is more "useful," "relevant" and appropriate for solving the most urgent problems of underdevelopment than the pursuit of science "for its own sake" according to the standards and the aims of the "international scientific community"?

It is only over the long term that science has an important role as a factor for educational, cultural and institutional change. In the medium term, and even more so in the short term, the capacity of a given country to innovate depends on its pool of technical skills and entrepreneurial talents rather than on producing a scientific elite. In the medium term, and even more so in the short term, there is no *economic* benefit to be derived from basic research. We laid strong emphasis in the book on the highly significant example of Japanese technical and industrial success, which at the beginning owed very little to basic research or even to the Japanese universities. The modernization and industrialization of Japan, like that of the newly industrialized countries, was not until recently accompanied by major contributions to scientific progress as such. The situation started to change in Japan because the very nature of its industrial development now requires a greater input of theoretical research. But how can one overlook the fact that this change is connected as much with the greater economic prosperity of the country as with the new prerequisites for producing technical innovations that are increasingly "sophisticated" and linked to laboratory work?

It should be noted in passing that the way in which statistics on research and development (R&D) are presented contributes—in the same way as the prestige attached to basic research in academic circles—to the distinctly questionable idea that there is a direct and proportional link between a country's scientific potential and its capacity to innovate. It is not because science is now the source of many technical innovations that it can be said that more science means a greater likelihood of producing innovations. In the R&D equation, the scientific part strictly defined is the smallest element; development is always the largest (about 60 per cent) in the industrialized countries, and in particular in those which have the greatest industrial "success." This is not simply because development is the most expensive part, it is also and above all because this phase of the research process is the most important before industrial production is embarked on. In fact, it is characteristic of R&D in the industrialized countries that it is performed largely by industry, with the aim of benefiting from scientific progress through applications and innovations with market potential. Conversely, the less developed a country, the poorer its industrial infrastructure, the less its research expenditures are directed at economic questions. The consequence is that in developing countries the share of university research is the largest element in R&D expenditures.

Hence our question: in most developing countries, where the university tradition and system are inadequate, what is the point of putting scarce resources (in brains, budgets and facilities) to work in poor conditions in order to extend or deepen the "scientific paradigm" when the most urgent problems could be solved using the available technologies more efficiently? Christian Comeliau emphasizes the difficulties that Third World countries would have if they gave up scientific research. Here again, we never said that they should, but how many developing countries produce new knowledge? If we stop talking about the Third World as if it were homogeneous, and look at the developing countries one by one, is it not obvious that most of them—whatever their efforts to create and expand their higher education system—are condemned to a mere *semblance* of scientific output?

It is true that the new technologies cannot be produced without an input of laboratory work. But, first, are the most advanced technologies really necessary for development? Secondly, do the exploitation, management and maintenance of these technologies need science as such? In the majority of these countries (we did not say *all* of them), what problems do they encounter if they do not devote too many resources to basic research or even none at all? Technological dependency? Surely the quickest way to reduce this dependency is to start instead by introducing universal primary education and training larger numbers of middle-level technicians.

We did not say that this dependency could not be reduced. Nothing in our view is forever, *provided* that these countries give themselves the means to change over the long term. We constantly challenged the determinist argument (see especially pp. 27–32) which Amilcar Herrera (1989) accuses us of adopting. Technical change is not *predestined,* we kept insisting. Must we point out again that at the beginning of this century, the economic and social conditions in many European countries were very similar to those in what we now call developing countries? Or that nowadays all the former Communist countries share most of the features of underdevelopment? In other words, development is not simply a relative concept, it is also subject to change: nobody ever wins, and *bad* policies can lead to retardation instead of progress. Are there not plenty of examples of such policy disasters in Europe as well as in the Third World?

The prerequisites are time, sustained effort and agreed priorities. Science is an end in itself for scientists, but it is simply one among many options in a development policy, and one that is certainly less useful in the medium term than technical know-how (linked as far as possible to a thorough scientific education). In short, as regards education, for the vast majority of developing countries that do not have a research infrastructure, a scientific tradition, an industrial system that can use the products of research, a highly trained workforce, and so on, the most urgent need is not to make a disproportionate investment in basic research, nor even to try to weaken the grip of technological dependency; it is to increase the skills that will enable them to overcome the limitations of their circumstances and spread practical knowledge of those technologies which, as they become more widespread and exploited, can contribute more efficiently to the development of the *whole* society.

Christian Comeliau rightly points out that our book does not deal directly with the social sciences. We concentrated on the problems raised by ways of handling "hard" sciences; it was certainly not our intention to challenge the role that the social sciences can play in gaining a better knowledge of—and therefore improving the chances of making one's own and mastering—the history, culture and workings of a society. Furthermore, we insisted several times that "technology transfer" does not mean that everything can be reduced to a package of "hardware," since any technological product transferred must be accompanied by "software," including the instructions for use, the technical standards, the management methods and the work organization, all in all an entire system of values—otherwise it becomes a "black box," and there is only very hazy knowledge of how it works and is maintained. We argued at even greater length that there is no difference in our view between transferring something like law—which is a "soft" discipline—and any technical product whose physical appearance suggests that its technology is based on "hard" sciences: imported regulations and laws (development law) are options of legal policy *to the same extent* as imported technologies are options of industrial policy (pp. 150–153). But we immediately added that if a country intends to benefit from those parts of the technical system necessary for its survival, it must have more technicians, supervisors and skilled workers than specialists in law (or sociology). Yet is it not still typical of developing countries (in Latin America and Africa) that they turn out far more teachers and researchers in social sciences than doctors, nurses, engineers and technicians?

Culture, economics and the rest

Of all those who made comments, Amilcar Herrera—while congratulating us on the soundness of our analysis—was the most critical, stressing above all his strong disagreement with our "characterization of the obstacles which hinder the adequate use of science and technology by the Third

World countries" (Herrera, 1989: 170). There are times when one must agree to disagree. Herrera offers a definition of modern science that we consider totally inadequate (and which, what is more, he misattributes to Auguste Comte). He tries to argue that there is no "discontinuity in the basic principles and attitudes that have guided the generation of scientific knowledge since the emergence of *Homo sapiens*" (Herrera, 1989: 171). Consequently "the intellectual foundations of modern science are not external to the cultural traditions of Third World countries" (ibid.).

> Science was recognized as an area of knowledge that could be clearly differentiated from philosophy only in the nineteenth century, with Auguste Comte's formulation: the study of natural, mental and social phenomena by empirical methods... The generation based on *empirical evidence* [our italics] has been essential for the survival and progress of mankind since the beginning of civilization (Herrera, 1989: 171)

Defined in this way, modern science is clearly not a Western invention, created by the "Greek miracle" or associated with the earliest history of the Mediterranean region (from Egypt via Judeo-Christian civilization to the Arabs), and all cultures and all societies ought to be ready *in the same way and to the same extent* to take on board this type of knowledge, insofar as it involves a kind of rationality similar to all the other kinds.

This is precisely what we would dispute, because scientific rationality seems to us (as indeed to most historians of science and epistemologists) to represent a *clean break* with the principles, the methods and the results of all the other kinds of knowledge. In the first place, for Auguste Comte, science was never reduced merely to "a study of phenomena by empirical methods." The "transition to the positivist era" involves not just the use of "empirical methods," but "explanation by laws" which are based equally on theory and proof. The study of nature through mathematics and the experimental method constitute a breach with empiricism—after all, our senses tell us that the Sun moves around the Earth, but scientific theory and experimental proof show us that the opposite is true. This modern science, whose heroes are Galileo, Descartes and Newton, relates to a quite different idea of the world.

For example, Joseph Needham has shown that the traditional Chinese idea of the world involves laws which do not establish a system of causal relationships: they are based on a sort of "organic co-operation" and they provide ways of understanding that are never *guaranteed* because these laws are subject to the humours, arbitrariness and the potential spite of the gods or the elements. The same can be said of the traditional culture of India, which still affects the thinking of the great majority of Indians: as Francis Zimmerman (1982: 217) has shown (and we quote his work several times, see especially our pp. 115–116), the underlying principle of ayurvedic science involves ritual, with techniques laid down by Revelation and Tradition, based on a conception of the world totally removed from any notion of a "natural science" created from a *search for causes*.

By contrast, modern science linked to the European cultural tradition *assumes* regular, permanent laws and this is what underpins its universality in practice. Einstein said "God is subtle, but he is not spiteful (*bösich*)." This kind of reasoning excludes the possibility of the laws of nature being caught wanting; at the same time, it is the foundation of a theoretical explanatory system with universal applicability. Let us take the example of the fight against smallpox. The first two phases of this fight did indeed involve "empirical evidence": first "variolization," consisting of inoculation with a mild case at a time when there was no epidemic (a method first used in China); then "jennerization," which involved inoculation with cowpox. The two methods arose from a simple observation that, in the first case, nobody catches smallpox twice; and in the second, cowmen and milkmaids rarely caught the disease. When the third stage was reached, with Pasteur and the development of a vaccine, it was not "empirical evidence" that determined the method, but the *theory* of infection by germs and the discovery that the infection could be weakened effectively by heating or dessication: the cause of the problem was identified and the remedy was universally efficacious. Yet the path that led Pasteur to this theory, from his early work on dissymmetry to the study of fermentations and microbes, was marked by discoveries and proofs that were far removed from "empirical evidence." Of course, this scientific reasoning can coexist with traditional thinking, for instance modern and traditional medicine in India or Japan; and we would stress that while traditional treatments sometimes work, it is impossible to offer any rational explanation for why they do and hence they cannot be turned into a principle that is universal. Exactly the opposite is true of "European" science. Shamanism, too, produces results, but what relationship is there between its practices and science as we know it?

It is not our intention to cast doubt upon these practices, but rather to demonstrate that they are based on a completely different intellectual system, far removed from experimental science. To highlight this fundamental difference *does not mean* that we are therefore arguing that the "European" scientific understanding of the world is the only one or the ruling one (especially in imperialist fashion). To quote Needham again, if Europe has made a great contribution to human history, it is indeed in the area of scientific reasoning underpinning the natural sciences, but that does not mean that the universal character of these branches of knowledge should lead to European civilization appearing to be the model for the whole world. The particular case of science is not a guarantee of the universality of the "European model"—as we have repeatedly stressed .

This is why it is impossible to minimize—whatever the weight of the strictly economic or political factors—the importance of cultural factors in possible explanations of the difficulties that many developing countries encounter in assimilating, disseminating and spreading the peculiar approach characteristic of Western science. We recalled what is self-evident: that economic and political dependence (in the past the power wielded by the colonial powers; nowadays the industrial domination of the "advanced" countries) was *one* of the factors which explain the "lag." But it is not the only one, and too great attention being paid to it carries a risk of falling into reductionist Marxist thinking. The *external* explanation cannot blithely ignore the other *internal* factors. Amilcar Herrera admits this to some degree himself, when he stresses that the main cause of underdevelopment is related to the "socioeconomic" strategies adopted in order to absorb technological innovations. How can cultural factors be kept out of these options?

On this matter, most of the examples we cite were originally given by representatives of developing countries themselves (pp. 74–78). From Nehru to Indira to Rajiv Gandhi, surely the lack of success surrounding the efforts to spread the "scientific temper" in India is a sign that "Western-style" rationalism has problems in catching on? We quote Thomas O. Eisemon on Nigeria and Kenya: "It would require much greater historical distance before offering a judgement on the success of introducing a scientific 'ethos' into Black Africa." We quote Abdul Rahman and Arnab Rai Choudhuri on India and A. B. Zahlan on the Arab world, all of them people with much experience of their regions, with the skills of historians of science and with international points of comparison. All their observations show that the cultural and institutional obstacles play an important and indisputable role in the difficulties encountered by some developing countries in taking to Western-style rationality. We do not say, however, that these problems are insurmountable, and we do not conclude (far from it) that attitudes, beliefs and behaviours which are not based on a system of logic impregnated with Western rationality are entirely without coherence, legitimacy and even a certain efficiency. We do conclude nevertheless that it takes time, a long time, and continuity in the policies on primary education for change to occur. In this regard, the most brutal assessment is not in the book (it is more recent, and it is so extreme that we might well not have used it). It is by an economist from Cameroon: "The unique cause, the one that is at the root of all the problems, is African culture, characterized by its self-sufficiency, its passivity, its lack of enthusiasm for going halfway to meet other cultures before they impose themselves and crush the local culture, its inability, once the damage has happened, to develop through contact with them without falling into contemptible mimicry" (Etounga-Manguelle, 1990).

In any case, our book stressed the importance of history, and so also of culture, in understanding the specific circumstances of "technological take-off" in the newly industrialized countries (or the increasing "lag" in the other developing countries) as against interpretations according to the "economic paradigm," whether of Left or Right, Marxist or liberal, which has always ignored the historical dimension. "Structuralism, the theory of dependency, monetarism or export strategies—none of them captures the variety and above all the inconsistencies of the development paths... The fact is that the world as it turns and societies as they evolve do not abide by the economists' models or the prophets' visions" (p. 135).

The reasonable, the desirable and the possible

The greatest lack of understanding (or the worst misinterpretation) of our book occurs when Amilcar Herrera, in wondering what the aims of development ought to be, reproaches us for advising developing countries "implicitly" to copy the model provided by the industrialized countries. But we *quite explicitly* repeat many times that there is no unique development model, and that the countries of the Third World should indeed avoid reproducing that of the industrialized countries (see in particular pp. 153–154). Development is a journey from tradition to modernity, and the only ones who should decide upon the ports of call (close to or far from their home port) are the countries that embarked on the voyage. "What should be retained from our traditions, what modern ways should we aim to adopt?" is the refrain throughout this journey. Some of the developing countries never set out, others had a bad start, and only a few have journeyed far. Did these last choose a form of development altogether similar to that of the most advanced countries? This would seem to us far from certain: none of the newly industrialized countries has completely abandoned its traditions or its culture. Each one has its special problems, which tend to be connected with its history and culture; and, since each country's circumstances are the product of a different past, each also creates its own story.

In this regard, technological independence is not only relative, it only makes sense if the choices made in order to achieve it match the real and balanced needs of each country. Against the prestige, the attraction and the promises of the "new technologies" we stress throughout the book the illusions and the costs of "blindly" copying the industrialized countries' experiences. In short, we cannot understand how Amilcar Herrera comes to have read what he has into our book, particularly since we agree with many of his comments, especially with his analysis of the situation, the dangers he sees and deplores in the widening gap between industrialized and developing countries, the need for solidarity and co-operation which are made even more essential as the rich, less rich and poor are all doomed to cope with the same challenges raised by the environmental problems affecting the whole Earth.

We have some difficulties in accepting Herrera's solutions since we feel that they are somewhat utopian in both their relevance and their plausibility. "The conscious limitation of our consumption of natural resources to a level compatible with their relative availability, and with the preservation of the global equilibrium of the biosphere" (p. 177) is certainly a desirable aim for the future of all mankind. But is that really the right response to the challenges of underdevelopment? It might *perhaps* be the right one if the developing countries were lagging behind *solely* because of the consumption patterns of the industrialized countries, an argument that we consider less than reasonable and that we cannot imagine to be Herrera's position. And while it is true that the problems of the environment on a global scale are now added to those of underdevelopment, might not the remedy he suggests bring about the exact opposite of the "desirable," i.e. greater selfishness and less solidarity on the part of the industrialized countries?

His remedy is essentially to prescribe "an austere material life" which would make it possible, with the help of technology, to achieve "the almost complete elimination of the social division of labour" (p. 177) and ensure the stability of an egalitarian society. To this end the industrialized countries should "renounce a social objective which in the Western conception of progress has been central since the consolidation of the Industrial Revolution: the indefinite increase in material welfare" (p. 179). As for the developing countries, "late-comers" to the current and supposedly universal notion of development, they have not yet taken on board the system of values on which this notion is based; and for most of them "'austerity' would mean a level of material well-being undreamed of in the present situation" (p. 179).

How should one react to this torrent of good intentions—which confuses the reasonable, the desirable, the possible and the actual—without appearing hardheadedly realistic, that is, incapable of giving in to utopian visions? "The fact that the proposed society is possible, desirable and necessary," writes Herrera, "does not mean that its implementation would be easy" (p. 178). There is no doubt about that: frugality is just as likely to be a popular remedy for economic problems as is sexual abstinence for overpopulation. The option of "material austerity" for mankind presupposes a different human species to the one whose mastery of nature has been founded on its scientific approach to problems. Yes, there is something promethean or faustlike in this approach, which has managed to turn knowledge into real power—for good or ill. Yes, this approach is reductionist, the world is "disenchanted" with it, and the knowledge and techniques that it makes possible *can* lead to disaster. But the knowledge and the techniques can *also* help to improve the lot of humanity. We do not believe that the adoption of austerity by the industrialized countries—even supposing that they were capable of it and were prepared to initiate it—would by itself bring about an improvement in the standards of living, working and survival in the developing countries. We tend to believe the contrary: without growth in the industrialized countries, where would it come from in the developing countries?

This does not mean, of course, that there is no need for a more intelligent management of both natural resources and the biosphere in the developing countries as well as in the industrialized countries. And we agree entirely with Amilcar Herrera when he writes "The root of the problem is not technology but the distribution of income" (p. 183). Did we not say just this in our conclusions? "To reconcile the ambitions for economic growth with greater distributive justice is doubtless the best way not only to avoid social upheavals but also, indeed, to ensure that growth continues. None of the technical solutions provided by science and technology can act as a substitute for this political and social imperative" (p. 150). Do we need to repeat that it is in the industrialized countries that greater wealth has eventually made possible a relatively fairer division? The smaller the cake, the less there is to be shared out. The society that Amilcar Herrera dreams of, which is frugal and egalitarian (inevitably frugal because egalitarian?), requires there to be either a spontaneous "conversion" of human nature or a political system that forces human nature to "change." Yet recent history suggests that one cannot change human beings as easily as economies.

Amilcar Herrera says that "the basis for the formulation of the R&D strategy should be *the scientific and technological demands of the proposed society*" (his italics). This approach implies a sequence of steps: definition of the character of the desirable society (p. 178). This begs the questions of *who* will define these requirements and decide *what* is "desirable," and *how*? Since the road to Hell is paved with good intentions, utopian visions (sometimes) point the way forward. But here we have good reasons for being sceptical: this century has witnessed the failure and the human cost of too many utopias in a position of power for these choices to be left to anything but straightforward democratic control.

Notes

1. The thesis has now been published (in 1991) by the University of Kentucky Press under the title *Scientists in the Third World*.

References

Comeliau, C. (1989). "Is the Third World Headed for Perpetual Dependency?", *Social Science Information* 28 (2): 431–4.

Etounga-Manguelle, D. (1990). *L'Afrique a-t-elle besoin d'un ajustement culturel?* Yaoundé: Editions les Nouvelles du Sud.

Gaillard, J. (1988). "La recherche scientifique en Afrique," *La Documentation française* 148(4): 3–30.

Gaillard, J. (1989). "Les chercheurs et l'émergence de communautés scientifiques nationales dans les pays en développement," Doctoral thesis, STS, Paris, Conservatoire National des Arts et Métiers.

Herrera, A. (1989). "The Advantages of Being a Late-Comer to What?" *Social Science Information* 28 (4): 823–40.

Salomon, J. J. and Lebeau, A. (1988). *L'écrivain public et l'ordinateur—Mirages du développement.* Paris: Hachette.

Zimmerman, F. (1982). *La jungle et le fumet des viandes. Un thème ecologique dans la médecine hindoue.* Paris: Gallimard/Le Seuil.

Notes

Introduction

1. Albert O. Hirschman, *Strategy of Economic Development* (New York, Norton & Co., 1978); originally published in French: *Stratégie du développement économique* (Paris, 1958).
2. Ibid., p. 144.
3. Ibid., pp. 154–155.
4. Albert O. Hirschman, *Essays in Trespassing: Economics to Politics and Beyond* (Cambridge, Cambridge University Press, 1981), p. 37.
5. Ibid., p. 1.
6. Jean-Jacques Servan-Schreiber, *Le défi mondial* (Paris, Fayard, 1980), p. 420.
7. Ibid., p. 373.
8. Ann Johnston and Albert Sasson (eds.), *New Technologies and Development* (Paris, UNESCO, 1986), introduction by Jean-Jacques Salomon.

Chapter 1: No Shortcut to Development After All

1. See the critical comments of François de Closets, *Tous ensemble: Pour en finir avec la syndicratie* (Paris, Le Seuil, 1985), pp. 111ff.
2. See Sylvie Brunel (ed.), *Asie-Afrique: Greniers vides, greniers pleins* (Paris, Economica, 1986).
3. For the data, see the annual reports of the World Bank, UNCED, and *Twenty-five Years of Development Co-operation* (Paris, OECD, 1985).
4. Jean Racine, "Autosuffisance alimentaire et transition nutritionnelle," in Brunel, op. cit. (n. 2), p. 47.
5. Max Weber, *Essays in Sociology*, edited by H. H. Gerth and C. Wright Mills (New York, Oxford University Press, 1970).
6. See Richard P. Suttmeier, *Research and Revolution: Science Policy and Societal Change in China* (Lexington, MA, Lexington Books, 1974); *Science and Technology in the People's Republic of China* (Paris, OECD, 1977); Jon Sigurdson, *Technology and Science in the People's Republic of China* (Oxford, Pergamon, 1980).
7. Joseph Needham, *The Grand Titration: Science and Society in East and West* (London, Allen & Unwin, 1969).

8. Jean-Jacques Salomon, *Le destin technologique* (Paris, Balland, 1992), p. 275.
9. Nathan Rosenberg, "Technology, economy, and values," in G. Bugliarello and D. B. Doner (eds.), *The History and Philosophy of Technology* (Chicago, University of Illinois Press, 1979), p. 84.

Chapter 2: The Many Third Worlds

1. Gunnar Myrdal, *The Challenge of World Poverty* (London, Allen Lane, 1970), p. 8.
2. Ibid.
3. W. W. Rostow, *The Stages of Economic Growth* (Cambridge, Cambridge University Press, 1960).
4. A. Gerschenkron, *Economic Backwardness in Historical Perspective* (Cambridge, Cambridge University Press, 1962).
5. S. N. Einsenstadt, *Modernization: Protest and Change* (Englewood Cliffs, NJ, Prentice-Hall, 1967).
6. Jacques Lesourne, *Les mille sentiers de l'avenir* (Paris, Seghers, 1981), p. 62.
7. *Facing the Future: Mastering the Probable and Managing the Unpredictable* (Paris, OECD, 1979); see also Jacques Giri, "Les tiers mondes face à eux-mêmes," in J. Lesourne and M. Godet (eds.), *La fin des habitudes* (Paris, Seghers, 1985).
8. Zbigniew Fallenbuchl, *East-West Technology Transfer: Study of Poland (1971–1980)* (Paris, OECD, 1983).
9. Myrdal, op. cit. (n. 1), p. 11.

Chapter 3: A Basic Discontinuity

1. Margaret Gowing, *Britain and Atomic Energy, 1939–45* (London, Macmillan, 1964).
2. Jean-Jacques Salomon, *Science and Politics* (Cambridge, MA, MIT Press, 1973).
3. See Fernand Braudel, *Civilisation matérielle, économie et capitalisme, XVe-XVIIe siècle*, Vol. 1 (Paris, Colin, 1979), p. 338.
4. Ibid., p. 345.
5. See "The titans of high technology—Japan and the United States: A survey," *The Economist* (August 25, 1986), pp. 5–20.
6. *OECD Science and Technology Indicators*, No. 2: *R&D, Invention and Competitiveness* (Paris, OECD, 1986).
7. *Rapport sur la compétitivité internationale*, World Economic Forum (Paris, Economica, 1985).
8. *Trade-related Issues*, Vol. 1: *The Semi-conductor Industry*; Vol. 2: *The Space Industry*; Vol. 3: *The Pharmaceutical Industry* (Paris, OECD, 1985).
9. *OECD Science Indicators: Resources Devoted to R&D* (Paris, OECD, 1984).
10. OCED, op. cit. (n. 6).
11. C. S. Hacklish, *International Joint Ventures* (Cambridge, MA, Lexington Books, 1985), and "Technical alliances in the semi-conductor industry" (New York University, Center for Science and Technology Policy, 1986); H. I. Fusfeld and C. S. Hacklish, "Cooperative R&D for competitors," *Harvard Business Review* (November–December 1985); D. C. Mowery, "Multinational joint ven-

tures in product development and manufacture: The case of commercial aircraft" (Pittsburgh, Carnegie Mellon University, Department of Social Sciences, 1986); B. Payne, *Co-operation and Technological Change in the Machine Tool Industry* (London, Technical Change Centre, 1986).

12. Seminar on Science and Technology Policy in Small Industrialized OECD Member Countries in Relation to Economic Growth, Helsinki, January 29–30, 1986.

13. *Report on the Differences in Technological Development Between the Member States of the European Community* (Brussels, European Parliament, 1985), Document A2-106.

14. Vivien Walsh, "Technology and competitiveness and the special problems of small countries," *STI Review* (Paris, OECD), No. 2 (1987), pp. 81–134.

15. Karl Marx, "General introduction to the critique of political economy, 1857," *Complete Works* (Harmondsworth, Pelican Marx Library, 1973).

16. Nancy Stepan, *Beginnings of Brazilian Science* (New York, Watson, 1981); Mario Guimaraes Ferri and Shozo Motoyama, *Historia das ciencias no Brasil*, 3 vols. (São Paulo, EDUSP/EPU, 1979–1981); Francisco Sagasti, *El desarrollo cientifico y tecnologico de America latina* (Mexico, Fundo de Cultura Economica, 1981).

17. *Science, Growth and Society* (Paris, OECD, 1971).

18. Joseph Hodara, "La conceptuacion del atraso cientifico-tecnico de America latina: El telon de fondo," *Commercio Exterior* (Mexico) (November 1976), pp. 1285–1291; see also Jorge Sabato (ed.), *El pensamiento Latinoamericano en la problematica ciencia-tecnologia-desarrollo-dependencia* (Buenos Aires, Editorial Paidos, 1975).

19. See Candido Mendès (ed.), *Le mythe du développement* (Paris, Le Seuil, 1977).

20. Amilcar Herrera, "Ressources naturelles, technologie et indépendance," in Mendès, op. cit. (n. 19), pp. 141–159; "Tecnologias cientificas y tradicionales en los paises en desarrollo," *Commercio Exterior* (Mexico), Vol. 28, No. 12 (December 1978), pp. 1462–1476; and "Endogenous generation of technology instead of imitated innovation," in E. U. von Weizäcker et al. (eds.), *New Frontiers in Technology Application* (Dublin, Tycooly, 1983).

21. Celso Furtado, *Le mythe du développement économique*, 2d ed. (Paris, Anthropos, 1984), p. 111.

22. Ibid., p. 112.

23. Ibid., p. 105.

24. Arghiri Emmanuel, *Technologie appropriée ou technologie sous-développée?* (Paris, Presses Universitaires de France, 1980).

Chapter 4: The Contemporary Technical System

1. Fernand Braudel, *Civilisation matérielle, économie et capitalisme, XVe-XVIIe siècle*, Vol. 1 (Paris, Colin, 1979), p. 291.

2. Bertrand Gille, *Histoire des techniques* (Paris, Gallimard, 1978), p. 19.

3. Joseph Schumpeter, "The *Communist Manifesto* in sociology and economics," *Journal of Political Economy*, Vol. 57 (June 1949), pp. 199–212.

4. Claude Lévi-Strauss, "Race et histoire," *Anthropologie structurale*, Vol. 2 (Paris, Plon, 1973).

5. See Maurice Daumas, *Histoire des techniques*, Vols. 2 and 3 (Paris, Presses Universitaires de France, 1965 and 1968), especially the Introductions.

6. Joseph Schumpeter, *Business Cycles*, Vol. 2 (New York, McGraw-Hill, 1939), pp. 226–227.

7. C. E. Shannon and W. Weaver, *The Mathematical Theory of Communication* (Urbana, University of Illinois Press, 1949).

8. Karl Popper, *The Logic of Scientific Discovery*, rev. ed. (London, 1972).

9. On the reasons why it is impossible to encourage technology push over and above market pull, see especially Nathan Rosenberg, *Perspectives on Technology* (Cambridge, Cambridge University Press, 1976), and *Inside the Black Box: Technology and Economics* (Cambridge, Cambridge University Press, 1982).

10. On the contributions of technology to scientific progress, see Maurice Daumas, *Les instruments scientifiques aux XVIIe et XVIIIe siècles* (Paris, Presses Universitaires de France, 1953).

11. Carnot's work was actually published so discreetly in 1824 that it made very little impact for ten years. The ideas in it—the basis of a new branch of science, thermodynamics—were not fully understood until 1846, when the future Lord Kelvin resurrected "Carnot's principle" from the little-known book; meanwhile, Rudolf Clausius discovered it independently.

12. John von Neumann, *Theory of Self-reproducing Automata*, edited by A. W. Burk (Urbana, University of Illinois Press, 1949).

13. Will the lessons drawn from research so far in artificial intelligence ever be reassuring? The most difficult human activities to reproduce are not those that are the most abstract and intellectual but those that involve feeling: intuition, perception, emotion. To be sure of winning, the computer playing chess, checkers or backgammon must not only have enormous computing power but must also learn to act and react in situations that cannot be reduced to algorithms; in other words, it must be able not only to simulate human knowledge but also human ignorance—as well as have the ability to take advantage of it.

14. See Patrick Lagadec, *La civilisation du risque* (Paris, Le Seuil, 1981), and *Le risque technologique majeur* (Paris, Pergamon, 1981). See also Jean-Jacques Salomon, *Le destin technologique* (Paris, Balland, 1992).

15. Paul A. David, "La moissoneuse et le robot: La diffusion des innovations fondées sur la micro-électronique," in J.-J. Salomon and G. Schméder (eds.), *Les enjeux du changement technologique* (Paris, Economica, 1986). Joseph F. Engelberger, the founder of the first company to produce industrial robots, Unimation Inc., which did not make a profit for the first ten years, is quoted as saying: "Nobody puts a robot to work because they want to make life easier for their employees. They put it to work for economic savings." Quoted in I. Asimov and K. A. Frenkel, *Robots: Machines in Man's Image* (New York, Harmony Books, 1986), p. 36.

16. Lynn M. Salerno, *Harvard Business Review* (November-December 1985).

Chapter 5: The Science of the Poor

1. Gilbert Etienne, "Tiers monde: Nouvelles stratégies du développement ou nouveau dogmatisme?" *Politique étrangère*, No. 4 (December 1981).

2. Bruno Latour, "Le centre et la périphérie: A propos du transfert des technologies" *Perspective et santé*, No. 24 (Winter 1982), and "Comment redistribuer le grand partage?" *Revue de synthèse*, No. 110 (April–June 1983), pp. 203–236.

3. See Lucien Lévy-Bruhl, *La mentalité primitive*, 1st ed. (Paris, Alcan, 1922).

4. Claire Salomon-Bayet, "Modern science and the coexistence of rationalities," *Diogenes*, No. 126 (April–June 1984), p. 12.

5. Joseph Needham, "Human law and the laws of Nature," in *The Grand Titration: Science and Society in East and West* (London, Allen & Unwin, 1969), p. 330.

6. See P. Huard, J. Bossy, and G. Mazars, *Les médecines de l'Asie* (Paris, Le Seuil, 1978); Fernand Meyer, *GSO-BA RIG-PA: Le système médical tibétain* (Paris, CNRS, 1983).

7. Abdul Rahman, *Triveni: Science, Democracy and Socialism* (Simla, Indian Institute of Advanced Study, 1977), p. 94.

8. See Martine Barrère, "La science en Inde," *La Recherche* (Paris), Vol. 17, No. 180 (September 1986).

9. Giovanni Rossi, "La science des pauvres," *La Recherche*, No. 30 (January 1973), pp. 7–14.

10. Jacques Gaillard, mission report on Senegal, International Science Foundation, Stockholm, 1985.

11. Jacques Gaillard, *Scientists in the Third World* (Lexington, University of Kentucky Press, 1991).

12. Derek J. de Solla Price, *The Difference Between Science and Technology* (Detroit, MI, Thomas Edison Foundation, 1968), and "Research on research," in David L. Arm (ed.), *Journeys in Science: Small Steps, Great Strides* (Albuquerque, University of New Mexico Press, 1967), pp. 10–11.

13. It was Georges Canguilhem who called the scientist a *"figure de la culture"*; see Jean-Jacques Salomon, *Science and Politics* (Cambridge, MA, MIT Press, 1973), p. 251.

14. Kapil Raj, "Hermeneutics and cross-cultural communication in science: The reception of Western scientific ideas in 19th-century India," *Revue de synthèse*, 4th series, Nos. 1–2 (January–June 1986), pp. 107–120.

15. Albert Szent-Györgi, *The Observer* (London), November 24, 1954.

16. Albert Szent-Györgi, *American Journal of Physics*, Vol. 43 (1975), quoted in Gerald Holton, *The Scientific Imagination: Case Studies* (Cambridge, Cambridge University Press, 1978).

17. Michael Moravcsik, *Science and Development: The Building of Science in Less Developed Countries* (Bloomington, University of Indiana Press, 1976).

18. John Garfield, "Mapping science in the Third World," *Science and Public Policy* (London) (June 1983); Beth Krevitt Eres, "Socioeconomic conditions relating to the level of information activity in the less developed countries," Ph.D. thesis, Drexel University, 1982.

19. Thomas Owen Eisemon, "The implantation of science in Nigeria and Kenya," *Minerva*, Vol. 12, No. 4 (1979), p. 526, and *The Science Profession in the Third World: Studies from India and Kenya* (New York, Praeger, 1982).

20. "Excellence in the Midst of Poverty," *Nature*, April 12, 1984.

21. V. Shiva and J. Bandyopadhyay, "The large and fragile community of scientists in India," *Minerva*, Vol. 18, No. 4 (1980), pp. 575–594.

22. A. B. Zahlan, *Science and Science Policy in the Arab World* (London, Croom Helm, 1980), p. 74.

23. Arnab Rai Choudhuri, "Practising Western science outside the West: Personal observations on the Indian scene," *Social Studies of Science* (London), Vol. 15 (1985), pp. 475–505.

24. Jean-Jacques Salomon, "Results of an OECD Survey," in *Training of Research Workers in the Medical Sciences* (Geneva, World Health Organization, 1973) pp. 97–105.

25. Charles Kidd, "Migration of medical researchers and doctors," in ibid.

26. For these statistics, see World Bank, *Report on World Development, 1983*.
27. Interview with Abdus Salam, "On threshold of superpower," *Indian Express*, May 29, 1981.
28. See Cheong Siew Yoong, "Making science, technology and mathematics education relevant to youth in developing countries," *Science and Public Policy*, Vol. 13 (June 1986), pp. 125–133.

Chapter 6: The Looking-Glass Race

1. Lewis Carroll, *Alice's Adventures in Wonderland* and *Through the Looking Glass*.
2. Francisco R. Sagasti and Cecilia Cook, "Tiempos dificiles: Ciencia y tecnologia en America latina durante el decenio de 1980" (Lima, GRADE, 1985), p. 75.
3. Philippe Chalmin and Jean-Louis Gombeaud, *Les marchés mondiaux en 1984–1985* (Paris, Economica, 1985).
4. For data, see Judith Sutz, "Informatique et société: Quelques réflexions à partir du tiers monde," Doctoral thesis, University of Paris I-Sorbonne (IEDES, 1984); "Quelle informatique pour le développement?" *Futuribles* (Paris) (September 1984); Armand Mattelart and Hector Schmucler, *L'ordinateur et le tiers monde: L'Amérique latine à l'heure des choix télématiques* (Paris, Maspero, 1983).
5. R. Saunders, J. Warford, and W. Bjorn, *Telecommunications and Economic Development* (Baltimore, MD, Johns Hopkins University Press, 1983).
6. International Institute of Communications, *The Use of Satellite Communication for Information Transfer* (Paris, UNESCO, 1982).
7. José Dion de Melo Teles, *Pela Valorizaçao da Intelogência* (Cadernos, Editora Universidade de Brasilia, 1985); Simon Schwartzman, "High technology vs. self-reliance: Brazil enters the Computer Age" (Rio de Janeiro, IUPERJ, May 1985), quoted in Antonio Botelho and Peter H. Smith (eds.), *The Computer Question in Brazil: High Technology in a Developing Country* (Cambridge, MA, MIT Center for International Study, 1985); Fabio Stefano Erber, "Microelectronics policy in Brazil," *ATAS Bulletin*, No. 2 (New York, United Nations CSTD, November 1985).
8. Dion de Melo Teles, op. cit. (n. 7), p. 116.
9. Zhang Xiaobin, "Microelectronic policy in China," *ATAS Bulletin*, No. 2 (New York, United Nations CSTD, November 1985), p. 129. The journal noted that the author, who for many years worked in the Chinese equivalent of a Western ministry of research and technology, had been appointed to head the first Chinese venture capital company, Venturetech Investment Corporation.

Chapter 7: The Machines from the North

1. Wassily Leontief, "The distribution of work and income," *Scientific American*, special issue on automation (September 1982), pp. 188–204.
2. Karl Marx, *Capital, a Critique of Political Economy,* ch. 15 (Harmondsworth, Pelican, 1976–1981). See also Alfred Sauvy, *La machine et le chômage: Le progrès technique et l'emploi* (Paris, Dunod, 1980).
3. Zvi Grilliches, "Le ralentissement de la productivité: La R-D est-elle la coupable?" in J.-J. Salomon and Geneviève Schméder (eds.), *Les enjeux du changement technologique* (Paris, Economica, 1986), p. 50.

4. See, for example, "Technology and employment," *STI Review* (Paris, OECD), No. 1 (Autumn 1986); *Technology and Structural Unemployment: Reemploying Displaced Adults* (Washington, DC, Office of Technology Assessment, 1986). There is a vast literature on the subject; see the survey in C. Freeman and L. Soete, *Information Technology and Employment: An Assessment* (Brighton, Science Policy Research Unit, University of Sussex, 1985).

5. *Technical Change and Economic Policy: Science and Technology in the New Socio-Economic Context* (Paris, OECD, 1980).

6. C. Freeman, J. Clark, and L. Soete, *Unemployment and Technical Innovation: A Study of Long Waves and Economic Development* (London, Frances Pinter, 1982); Christopher Freeman and Luc Soete, *Technical Change and Full Employment* (Oxford, Basil Blackwell, 1986).

7. Marc Uri Porat, "The Information Economy," Ph.D. thesis, University of Michigan (1976) (published as OT Special Publication 77-12, Washington, DC, Department of Commerce, 1977).

8. See Freeman, Clark, and Soete, op. cit. (n. 6); C. Freeman and L. Soete (eds.), *Technical Change and Full Employment* (Oxford, Basil Blackwell, 1987); N. M. Baily and A. K. Chakrabarti, *Innovation and the Productivity Crisis* (Washington, DC, Brookings Institution, 1988).

9. See Geneviève Schméder, "L'innovation est-elle à l'origine des cycles économiques?" *La Recherche, supplément Economie* (Paris), No. 183 (December 1986), pp. 27–31; N. Rosenberg and C. R. Frischtak, "Technological innovation and long waves," in C. Freeman (ed.), *Design, Innovation and Long Waves in Economic Development* (London, Design Research Publications, 1984).

10. Leontief, op. cit. (n. 1), p. 188.

11. Milton Santos, *L'espace partagé: Les deux circuits de l'économie urbaine des pays sous-développés* (Paris, Editions M.-Th. Génin, 1975).

12. Wassily Leontief and Faye Duchin, *The Future Impact of Automation on Workers* (New York, Oxford University Press, 1986).

13. Ibid., p. 25.

14. Harley Shaiken, *Work Transformed: Automation and Labor in the Computer Age* (New York, Holt Rinehart, 1985), p. 32.

15. Leontief, op. cit. (n. 1), p. 192.

Chapter 8: The Cathedrals in the Desert

1. Francis Zimmerman, *La jungle et le fumet des viandes: Un thème écologique dans la médecine hindoue* (Paris, Gallimard, Le Seuil, 1982), p. 217.

2. Ibid., p. 174.

3. Claire Salomon-Bayet, *Pasteur et le révolution pastorienne* (Paris, Payot, 1986), p. 18.

4. Elisabeth L. Einsenstein, *The Printing Press as an Agent of Change* (Cambridge, Cambridge University Press, 1980), p. 703.

5. Joseph Weinzenbaum, *Computer Power and Human Reason* (San Francisco, W. H. Freeman, 1976).

6. Final declaration of the international conference on cultural imperialism, Algiers, October 11–15, 1977; quoted in Jacques Perrin, *Les transferts de technologie* (Paris, Maspero, 1983), p. 107. See also Henri Pigeat, *Le nouveau désordre mondial de l'information* (Paris, Hachette, 1987).

7. Bernadette Madeuf, *L'ordre technologique international: Production et transferts* (Paris, La Documentation Française, 1981).

8. Stephen Hill, "Eighteen cases of technology transfer to Asia/Pacific

Region countries," *Science and Public Policy*, Vol. 13, No. 3 (June 1986), pp. 162–169.

9. A summary of Coutouzis's doctoral thesis (for the Université de Paris-Dauphine, January 1984) can be found in Mickès Coutouzis and Bruno Latour, "Le village solaire de Frangocastello: Vers une ethnographie des techniques contemporaines," *L'Année Sociologique* (Paris), No. 36 (1986).

10. Ibid., p. 163.

11. Claude Courlet and Pierre Judet, "Industrialisation et développement: La crise des paradigmes," *Revue Tiers Monde* (Paris), No. 107 (July–September 1986), p. 529. The study mentioned was carried out by the Institut de l'Entreprise (Centre Nord-Sud), *Pour un vrai partenariat industriel avec l'Afrique: Bilan et perspectives de l'industrie africaine* (Paris, 1985).

12. See the evaluation by Michel Chatelus, "Revenus pétroliers et développement: Leçons de l'expérience arabe," *Revue Tiers Monde* (Paris), No. 107 (July–September 1986), pp. 659–668.

13. For all these figures, see the following annual publications: *World Armament and Disarmament, SIPRI Yearbook* (Stockholm, SIPRI); *The Military Balance* (London, International Institute for Strategic Studies); *Ramses* (Paris, Institut Français des Relations Internationales).

14. Gabriel Palma, "Dependency: A formal theory of underdevelopment or a methodology for the analysis of concrete situations of underdevelopment?" *World Development*, Vol. 6, Nos. 7–8 (1978).

15. David S. Landes, *The Unbound Prometheus: Technological Change and Industrial Development in Western Europe from 1750 to the Present* (Cambridge, Cambridge University Press, 1969), p. 538.

Chapter 9: The Newly Industrialized Countries

1. Joseph Needham, *Science and Civilisation in China*, Vol. 1 (Cambridge, Cambridge University Press, 1954), p. 149.

2. René Etiemble, *Les Jésuites en Chine: La querelle des rites (1552–1773)* (Paris, Julliard, 1966); Jonathan D. Spence, *The Memory Palace of Matteo Ricci* (New York, Viking, 1983).

3. Hu Shih, *The Chinese Renaissance* (Chicago, University of Chicago Press, 1934), quoted by Needham, op. cit. (n. 1), pp. 146–147.

4. Max Weber, *The Protestant Ethic and the Spirit of Capitalism*, trans. by T. Parsons (London, Allen & Unwin, 1965).

5. Etienne Balazs, *La bureaucratie céleste* (Paris, Gallimard, 1968), pp. 311–312.

6. David S. Landes, *The Unbound Prometheus: Technological Change and Industrial Development in Western Europe from 1750 to the Present* (Cambridge, Cambridge University Press, 1969), p. 545.

7. A. Liepietz, *Mirages et miracles: Problèmes de l'industrialisation dans le tiers monde* (Paris, La Découverte, 1985), pp. 5–6.

8. A. Fishlow, "The evolution of economic theories in Latin America," in Interamerican Development Bank (IDB), *Report* (Washington, IDB, 1985), ch. 5.

9. Claude Courlet and Pierre Judet, "Industrialisation et développement: La crise des paradigmes," *Revue Tiers Monde* (Paris), No. 107 (July–September 1986), p. 526.

10. The nine countries covered by these statistics are Australia, Canada, France, Germany, Italy, Japan, Sweden, United Kingdom, United States. In the first period (1965–1973), only Australia and Japan had growth rates above 20

percent. For all these data, see *The New Industrializing Countries: Challenge and Opportunity for OECD Member Industries* (Paris, OECD, 1988).

11. Elisabeth Cattapan-Reuter, "L'industrie à l'époque de l'Encilhamento," in Frédéric Mauro (ed.), *La Préindustrialisation du Brésil: Essais sur une économie en transition* (Paris, Editions du CNRS, 1974), p. 64.

12. Stefan Zweig, *Brazil: Land of the Future*, trans. by A. St. James (London, 1942).

13. Renato P. Dagnino, "The arms industry in Brazil: The role of the state in technological development," contribution to the Proyecto Prospectiva Technológica para America Latina supported by the United Nations University (University of Campinas, 1985).

14. Ibid.

15. Paulo Bastos Tigre, "Performance and perspectives of the Brazilian computer industry," and Peter Evans, "Varieties of nationalism: The politics of the Brazilian industry," in A. Botelho and P. H. Smith (eds.), *The Computer Question in Brazil: High Technology in a Developing Country* (Cambridge, MA, MIT Center for International Studies, 1985).

16. Simon Schwartzman, "High technology versus self-reliance: Brazil enters the Computer Age," in ibid., p. 30.

17. On South Korea, see Mario Lanzarotti, "L'industrialisation en Corée du Sud: Une analyse en sections productives," *Revue Tiers Monde*, No. 107 (July–September 1986), pp. 639–657; Léon Vandermeersch, *Le nouveau monde sinisé* (Paris, Presses Universitaires de France, 1986), esp. chs. 1–2; Jacques Perrin, *Les transferts de technologies* (Paris, Maspero, 1983), pp. 72–78.

18. Joan Robinson, *Aspects of Development and Underdevelopment* (Cambridge, Cambridge University Press, 1979).

19. Michael Polanyi, "The Republic of Science: Its political and economic theory," in E. Shils (ed.), *Criteria for Scientific Development: Public Policy and National Goals* (Cambridge, MA, MIT Press, 1968), pp. 1–20; Polanyi, *The Logic of Liberty* (London, Routledge and Kegan Paul, 1951); Karl Popper, *The Open Society and Its Enemies*, 2 vols., 5th ed. (London, Routledge and Kegan Paul, 1966).

20. Patrice de Beer, "La Chine au grand vent des réformes économiques", *Le Monde diplomatique* (Paris), No. 388 (July 1986), p. 9. There is a vast literature on these reforms, especially their implications for science and technology. An Indian study based on intensive field work is especially interesting; see V. P. Kharbanda and M. A. Qureshi, *Science, Technology and Economic Development in China* (New Delhi, Navrang, 1987). See also the special issue of *China Report: A Journal of East Asian Studies* (Delhi), No. 12-1 (January–March 1986).

21. Albert Hirschman, *The Passions and the Interests: Political Arguments for Capitalism Before Its Triumph* (Princeton, NJ, Princeton University Press, 1977).

22. Estimate according to Kharbanda and Qureshi, op. cit. (n. 20).

23. Huang Zhizi, "China's challenges in changing times," *China Daily* (April 8, 1986), p. 4.

Chapter 10: History's Revenge

1. Louis Emmerij, "The share of income among nations," *OECD Observer*, No. 143 (November 1986), p. 11 (our italics).

2. Albert O. Hirschman, *Strategy of Economic Development* (New Haven, CT, Yale University Press, 1959).

3. Albert O. Hirschman, "The changing tolerance for income inequality in the course of economic development," in his *Essays in Trespassing: Economics to*

Politics and Beyond (Cambridge, Cambridge University Press, 1981), ch. 3.
   4. In Albert O. Hirschman, "A dissenter's confession: *The Strategy of Economic Development* Revisited," in Gerald Meier and Dudley Speers (eds.), *Pioneers of Economic Development* (New York, Oxford University Press, 1984), p. 108.

> Along with my fellow pioneers, I thus stand convicted of not having paid enough attention to the political implications of the economic development theories we propounded. But perhaps it was not altogether unfortunate that we were myopic and parochial. Had we been more far-sighted and interdisciplinary we might have recoiled from advocating any action whatever, for fear of all the lurking dangers and threatening disasters.

   5. Hirschman, op. cit. (n. 3), p. 59.
   6. See Roger Granger, "Droit du développement économique et social," *Encyclopaedia Universalis*, Vol. 6, pp. 42–46.
   7. See Joseph Fayet, *La révolution française et la science* (Paris, Marcel Rivière, 1960); Charles C. Gillespie, *Science and Policy in France at the End of the Old Régime* (Princeton, NJ, Princeton University Press, 1980).
   8. Nicolas Jéquier, *Appropriate Technology: Problems and Promises* (Paris, OECD, 1976); Nicolas Jéquier and Gérard Blanc, *The World of Appropriate Technology: A Quantitative Analysis* (Paris, OECD, 1983).
   9. E. Schumacher, *Small Is Beautiful* (London, Abacus, 1973).
   10. On appropriate technologies in practice, see Serge Michailof, *Les apprentis sorciers du développement* (Paris, Economica, 1984), pp. 129–140, and his more detailed study of the profitability of the *khandsari* industry in Christine Brochet (ed.), "Technologies appropriées et industrialisation," mimeo report (Paris, Ministry of Cooperation and Development, 1981), pp. 155–190.
   11. Jean Gimpel, "Le Moyen Age au service du tiers monde," in Régine Pernoud, Jean Gimpel, and Raymond Delatouche, *Le Moyen Age pour quoi faire?* (Paris, Stock, 1986).

Index

AEROTEC, 139
Africa, 10, 11, 13, 29, 93, 125. *See also specific countries*
Agence France Presse, 120
Agrarian reform, 142-143, 183
Agriculture, 13, 27, 30, 69, 80, 104, 107, 126, 154; Chinese, 18, 142-143; and population growth, 10-11
Aguilar, Alonso, 42
Algeria, 11, 29, 30, 93, 126, 127, 149
Aluminum production, 30
Americanization, 16
Arab countries, 15, 76-77, 126. *See also specific countries*
Arab League, 27
Argentina, 28, 29, 43, 46, 75, 93, 121, 128
Arms. *See* Weapons
Asia, 10, 11, 93, 136; development assistance programs in, 122-123; poverty in, 13, 29. *See also specific countries*
Associated Press, 120
AT&T, 39
Austerity, 177, 198-199
Australia, 39, 136
Austria, 39, 99
Automation, 61, 113
Automobile industry, 37

Babbha, H. J., 76
Bahrain, 126

Balazs, Etienne, 134
Belgium, 39
Bell, Daniel, 103
Bengal, 72
Bernardini, 128
Biomedicine, 42, 69
Biosphere. *See* Environment
Boeing, 39
Bolivia, 30
Bose, S. N., 76
Brain drain: impact of, 78-81
Brandt Report, 170
Braudel, Fernand, 49
Brazil, 6, 28, 29, 46, 121, 136, 137, 187; arms production in, 128, 139; computer industry in, 93, 94-97, 140-141; development in, 30, 45, 138-140; economy in, 148-149; GDP in, 11, 12; nuclear program in, 43-44; research in, 86, 142, 166-167
Brillouin, L., 53
Bull, 96
Burma, 10. *See* Myanmar
Burroughs, 96
"Buy America" Act, 95
Byron, Lord, 99-100

California Institute of Technology (Caltech), 39
Caltech. *See* California Institute of Technology

Canada, 39, 80, 136
Capital, 5, 34
Capitalism, 48, 173, 188; cultural context of, 144-145; industrial, 26, 50-51; and technology, 134-135
Carnot, Nicolas: *Réflexions sur la puissance motrice du feu,* 56
Cassini brothers, 78
Catholic University of Rio de Janeiro, 94-95
Chalmin, Philippe, 87
Chile, 30, 46, 75, 86, 149
China, 6, 10, 11, 30, 44, 67, 94, 97, 124, 128, 134, 137, 138, 149; missionaries in, 131-132; modernization in, 142-146, 153; research in, 166-167; science in, 17-18, 132-133, 194
Choudhuri, Arnab Rai, 196
Christianity, 131
CII, 93
Club of Rome, 46
COBRA. *See* Computadores Brasilerias
Colombia, 86, 149
Combine-harvesters, 60
Communism, 145
Competition, 10, 40, 100; in computer industry, 96-97; industrial, 35-36; in information technologies, 91-92
Computadores Brasilerias (COBRA), 95
Computer industry, 2, 37, 91, 100, 105, 117; in Brazil, 140-141; data bases, 118-119; development of, 92-93, 94-97; and telecommunications, 93-94
Computerization, 112
Comte, Auguste, 138, 171, 194
Confucianism, 145, 146
Cooperation, 5, 6, 39-40, 43-44
Copper production, 30
Costa, João da, 63
Côte d'Ivoire, 29, 30
Courlet, Claude, 125
Coutouzis, Mickès, 123
Crécy, battle of, 34
Crete, 123-124
Crozier, Michel, 152

Cruz, Oswaldo, 42
Cultural Revolution, 17, 18, 67, 143, 153
Culture: information technologies and, 118-119; and knowledge, 171-172; legal systems and, 152-153; and science, 132-133, 194-196; science and technology and, 18-19, 20, 21, 196-197

Dagnino, Renato, 139
Das Kapital (Marx), 24
Data bases, 118-119
David, Paul, 60
Debt, 25
Defense, 39, 127-128
Deindustrialization, 108
Delhi, 76
Democracy, 142, 144
Deng Xiaoping, 17, 142, 143, 144, 145
Dependencia, 44, 148
Dependency, 92, 118, 166, 196; and information technology, 94, 98; technological, 129, 181-182, 192
Descartes, René, 78
Developing countries, 25, 47, 114, 121, 135; defining, 23-24; differences among, 28-29; economies of, 2, 147-148; food production in, 10-11; GDP in, 11-12; information technologies in, 110-111; R&D in, 69-71. *See also* Third World
Development, 6, 13, 85, 153, 177, 179; in China, 142-146; economics of, 2, 3-4; factors in, 29-31; in Iran, 15-16; in Japan, 14-15; models of, 24, 175-176, 197; programs for, 122-124; science and technology and, 5, 186-187; strategies for, 156-158; theories of, 24-25, 210(n4)
Digibras, 95
Drucker, Peter, 103
Dupont, 39

EC. *See* European Community
ECLA. *See* United Nations Economic Commission for Latin America

Ecological policy: global, 177-178
Economic growth, 11, 46, 107, 147, 149, 175, 208-209(n10); and manufacturing, 12-13
Economics, 3-4, 24, 31, 135
Economies, 29; of developing countries, 2, 147-148; industrialization and, 101-102; raw materials and, 87-88; technology and, 47-48
Ecuador, 11, 86
Edison, Thomas, 78
Education, 1-2, 104, 105, 136, 154; role of, 190-191; and science, 80-81; and technology, 19, 20, 158; technology transfer and, 21-22
Egypt, 11, 93, 126, 127
Ehrlich, A.: *The Population Bomb*, 170
Ehrlich, D.: *The Population Bomb*, 170
Einstein, Albert, 66
Eisemon, Thomas O., 75, 196
Eisenstadt, S. N., 25
Eisenstein, Elizabeth, 117
Electronics industry, 37, 38
Elite: scientific, 41-42, 72, 80-81
EMBRAER, 128, 139
Emigration: of scientists, 78-81
Emmanuel, Arghiri, 47-48
Emmerij, Louis, 147
Empiricism, 55
Employment, 164; industrialization and, 99-100; Information Revolution and, 101-102; information technologies and, 103-104, 108-110; in service sector, 107-108, 112; technology and, 106-107
Energy, 45, 53, 123-124
ENGE, 139
Engineering, 39
England, 34. *See also* United Kingdom
Environment, 46, 176, 177, 199
Environmentalism, 21
ESPRIT, 39
Ethiopia, 11, 30
Eureka, 39
Europe, 33, 34, 40, 101, 108. *See also specific countries*
European Community (EC), 39
"Excellence in the Midst of Poverty," 76
Exports, 28, 87, 126, 136

Far East, 29. *See also specific countries*
Ferranti, 95
Field Medal, 71
Flanders, 34
Food production, 10-11, 12, 13
Food subsidies, 126, 127
Ford, 39
France, 9, 34, 37, 38, 39, 120, 153; computer industry in, 93, 95, 96
Freeman, Christopher, 101
Fundamentalism, 16, 17, 18
Furtado, Celso, 47, 138

Galileo Galilei, 78
Gandhi, Indira, 68, 196
Gandhi, Rajiv, 68, 80, 196
GATT. *See* General Agreement on Tariffs and Trade
General Agreement on Tariffs and Trade (GATT), 37, 95
General Electric, 39
General Motors, 39
Germany, 24, 37, 38, 39
Gerschenkron, A., 24
Ghana, 11
Gilles, Bertrand, 49
Gimpel, Jean, 156
Global Challenge, The (Servan-Schreiber), 4
Gombeaud, Jean-Louis, 87
GRADE, 85-86
Great Britain. *See* United Kingdom
Great Leap Forward, 17, 18, 143, 153
Greece, 40, 123-124
Group of 77, 27, 63
Guinea, 30
Gulf states, 30, 126-127. *See also specific countries*
Gunpowder, 34-35

Health services, 67

Herrera, Amilcar, 47, 169-184
High tech: defining, 36-37; R&D in, 38-40; trade in, 37-38
Hill, Stephen, 122
Hirschman, Albert O., 2, 3, 4, 32, 144, 149, 151
Hodara, Joseph, 44
Homo faber, 49, 189
Hong Kong, 12, 28, 29, 67, 97, 136
Huguenots, 78
Hu Shih, 132-133
Huygens, Christian, 55-56

IBM, 39, 93, 95, 96
ICL, 93
IEA. *See* International Energy Agency
IMBEL, 128
Imperial Academy of Sciences (Japan), 14
Imperialism, 9
Imports, 136
Import substitution, 42
Income, 157; distribution of, 150, 173-174, 183, 199; per capita, 11, 29
India, 6, 10, 11, 13, 28, 44, 67, 80, 94, 97, 115-116, 128, 137; alternative technology in, 155-156; development in, 30, 141; as NIC, 136, 187; research in, 69, 166-167; science in, 68, 72, 194-195, 196; working conditions in, 75-76
Indonesia, 10, 11, 30, 97, 136
Industrialization, 3, 9, 13, 14, 24, 45, 47, 50, 103, 121, 129, 137, 149, 157, 187; economies and, 101-102; and employment, 99-100, 111; impacts of, 25, 29-30; Iran, 15-16; Latin America, 42-43; Third World, 27-28, 159
Industrial Revolution, 24, 25-26, 52, 99, 108, 198; education and, 190-191; impacts of, 33-34; information technologies of, 53, 57; technical system of, 50-51
Industry, 12, 18, 89, 97, 138, 192; competition in, 35-36; high-tech, 36-40; linkages in, 157-158; projects, 125-126; and technology, 2-3, 19-20
Infant mortality, 12
Information, 53, 117, 125; control of, 119-121; employment and, 103-104; need for, 58-59
Information Economy, The (Porat), 103
Information Revolution, 2, 4, 28, 92, 103, 104, 114; impacts of, 99, 101, 102, 116-117
Information technologies, 4-5, 57, 118; and culture, 118-119; in developing countries, 110-111; development of, 92-98; employment and, 106-107, 108-110; impacts of, 86-87, 102-103, 104-105; politics of, 119-120; production of, 88-92
Innovation, 19, 89; and competition, 40, 92; impacts of, 180-181; as process, 20-21; and R&D, 191-192
Institute of Aeronautical Technology (Brazil), 138-139
Institute of Theoretical Physics (Madras), 76
Interamerican Development Bank, 135
Interfutures (OECD), 170
Intermediate Technology Group, 155
International Energy Agency (IEA), 123
International Telecommunications Union (ITU), 119
Interventionism, 140; impacts of, 141-146
Invention, 52
Investment, 19, 27, 31, 36, 86, 126, 138, 157
Iran, 18, 93; economic development in, 15-16; revolution in, 16-17
Iraq, 11
Ireland, 40
Irrigation, 124
Italy, 39
ITT, 39
ITU. *See* International Telecommunications Union

Jacquard, Joseph, 54
Jamaica, 30
Japan, 67, 88, 94, 95, 107, 108, 120, 136, 191; development in, 13-14, 24, 33; electronics industry in, 37, 38; production in, 91-92, 101; R&D in, 38, 39, 142; technology in, 74, 172
Jesuits, 131, 132
Judet, Pierre, 125

K'anghi, 132
Kanpur, 76
Kastler, Alfred, 189
Kenya, 29, 30, 75, 155-156
Khandsari plants, 155-156
Kharagpur, 76
Khomeini, Ruhollah, 16
Kidd, Charles, 79
Knowledge, 26, 59, 115; and culture, 171-172; and Information Revolution, 116-117; scientific, 5, 13, 55-56, 131, 165; and technology transfer, 124-125
Kodak, 39
Krishnan, M. S., 76
Kyodo, 120

Labor, 59, 91, 126-127, 154-155
Landes, David, 129
Latin America, 2, 11, 29, 94, 135; computer industry in, 92, 93; income distribution in, 173-174; R&D in, 85-86; science and technology in, 41-46. *See also specific countries*
Legal systems, 124-125, 152-153
Leisure, 113
Leontief, Wassily, 99, 101, 102, 110, 113-114
Lévi-Strauss, Claude, 52
Liberalization, 143-144
Libya, 11
Limits to Growth (Meadows and others), 46, 170
Lucretius, 66
Ludd, Ned, 100

Madras, 76

Malaysia, 11, 29, 30, 97, 136
Malnutrition, 12
Manufacturing, 39, 40, 57, 107, 125, 136; in Arab countries, 126, 128; in Brazil, 138, 148-149; and economic growth, 12-13, 28
Maoism, 146
Mao Zedong, 142, 153
Marx, Karl, 50; *Das Kapital,* 24
Marxism, 48
Massachusetts Institute of Technology (MIT), 39, 46
Masuda, Yoneji, 103
Meadows, D. H.: *Limits to Growth,* 170
Mechanization, 50, 51, 100, 113, 156; and automation, 60-61; and employment, 107, 109-110
Medicine, 67, 116, 195
Meiji restoration, 13-14, 24, 142
Mexico, 11, 28, 29, 45, 46, 94, 121, 136; computer industry in, 93, 97
Microelectronics, 58, 112, 119
Middle East, 13, 93. *See also* Arab countries; *specific countries*
Military, 34-35, 38, 139
Missionaries, 131-132
MIT. *See* Massachusetts Institute of Technology
Modernization, 21, 25, 47, 150, 169, 186, 191; and capitalism, 173, 188; in China, 142-146, 153; in Latin America, 42-43, 44
Moravcsik, Michael, 74
MOTORTEC, 139
Multinational firms, 6, 28, 38, 45, 47, 95, 128, 138; technology transfer by, 121-122
Myanmar, 10
Myrdal, Gunnar, 23, 24, 32, 187

Namibia, 93
NASA. *See* National Aeronautics and Space Administration
National Aeronautics and Space Administration (NASA), 38
Nationalism, 96, 139, 148, 179

National Science Foundation, 36
Nation-state, 5
Needham, Joseph, 66, 131, 132, 133, 194
Nehru, Jawaharlal, 68, 196
Nepal, 156
Netherlands, 39
Neumann, John von, 53, 60
New Development Strategy (Poland), 31
Newly industrialized countries (NIC), 12, 28, 85, 148, 186; characteristics of, 136-138, 187-188; and OECD, 135-136
New world order, 179-180
NICs. *See* Newly industrialized countries
Nigeria, 11-12, 29, 93
Nobel Prizes, 71
NOMIC, 119-120
Nonaligned nations, 27
North-South issues, 119-120, 169, 178
Norway, 39
Nuclear power, 21, 69; development of, 17-18, 35; impacts of, 34, 176; Latin America, 43-44

OECD. *See* Organization for Economic Cooperation and Development
Office of Science and Technology for Development, 64
Oil production, 87; and development, 27, 29, 30, 45; and economies, 11-12, 15, 126-127
OPEC, 27, 30, 87
Organization for American States, 44
Organization for Economic Cooperation and Development (OECD), 12, 36, 101, 123, 148; *Interfutures,* 170; newly industrialized countries and, 135-136; R&D expenditures and, 38-39
Ottoman Empire, 187

Pakistan, 10, 28, 29, 30, 155
Paraguay, 30
Patents, 69, 145

Pernoud, Régine, 156
Peru, 30, 75, 86
Pharmaceuticals, 37
Philippines, 10, 28, 29, 155
Philip VI, 34
Plan Calcul, 96
Planning, 31, 125-126
Plato, 78
Poland, 31
Polanyi, Michael, 142
Politics, 24, 31; and science, 44-46
Popper, Karl, 54, 142
Population Bomb, The (Ehrlich and Ehrlich), 170
Population growth, 10-11, 13, 30
Porat, Marc Uri: *The Information Economy,* 103
Portugal, 40
Positivism, 138
Postindustrial age, 103
Poverty, 9, 12, 13, 29, 31, 46, 149, 175, 176
Presidential Report on the Year 2000, 170
Press agencies, 120
Printing, 117-118, 187
Production, 3, 12, 121; information technologies, 88-92; and raw materials, 87-88; systems of, 59, 60-61
Productivity, 109
Protectionism, 95, 96, 140
Public sector, 126
Qatar, 126

Rahman, Abdul, 68, 196
Raman, C. V., 76
R&D. *See* Research and development
Rationalism, 17, 66
Raw materials, 87-88
Réflexions sur la puissance motrice du feu (Carnot), 56
Regulation, 20, 71, 74-75
Renaissance, 133
Research, 26, 38, 64, 164, 204(n13); applied, 55, 56-57, 71-72; basic, 73-74, 77-78, 183, 188-190, 191; and

politics, 141-142; scientific, 19, 67-68, 158, 164, 166-167; at universities, 72-73, 139-140
Research and development (R&D), 36, 128, 173-174, 181, 199; cooperation in, 39-40; expenditures on, 38-39, 41, 69-71; and innovation, 191-192; in Latin America, 85-86; policies for, 181-182; and regulations, 74-75
Resources, 27, 46, 56; changes in, 87-88; natural, 30-31, 177, 199
Reuters, 120
Revolution, 16-17, 51-52, 115, 142, 143
Ricardo, David, 109
Ricci, Matteo, 131, 146
Robinson, Joan, 141, 142
Robots, 60, 61, 112-113, 204
Rossi, Giovanni: "The Science of the Poor," 68
Rostow, W. W., 24
Rural sector, 13, 80, 104, 111
Russia, 24, 120

Sagasti, Francisco, 85-86
Saha, M. N., 76
Saha Institute, 76
Sahel, 11
Salam, Abdus, 80
Salomon-Bayet, Claire, 116
Samurai, 14
Santos, Milton, 111
São Paulo, 95
Saouma, Edouard, 10
Sarney, José, 47
Saudi Arabia, 12, 30, 126
Schumacher, E. F.: *Small Is Beautiful*, 155
Schumpeter, Joseph, 50, 52, 108
Schwartzman, Simon, 141
Science, 14, 17, 26, 115-116, 131, 138, 154, 188; in developing countries, 68-69; and development, 5, 186-187, 190-192; as elitist, 41-42, 80-81; European, 133-134; in Latin America, 42-46; and modernization, 149-150; production of, 165-166; role of, 190, 192-193; social, 150-151, 193; social and cultural factors in, 18-19, 67-68, 171-172, 181, 194-196; and technology, 55-56, 158, 189; in Third World, 77-78, 193-194; in traditional societies, 132-133; Western, 54, 66-67, 163-164, 168; working conditions and, 74-76
"Science of the Poor, The" (Rossi), 68
Scientists, 73; emigration of, 78-81; international community of, 65-66; North-South divisions of, 64-65; working conditions of, 74-77
Semiconductors, 38
Senegal, 71
SERPRO, 96
Servan-Schreiber, Jean-Jacques: *The Global Challenge*, 4
Service sector, 12, 13, 102-103, 107-108, 112
Seventh Plan, 80
Shannon, C. E., 53
Singapore, 12, 28, 29, 97, 136, 137, 142
Skills, 26, 78
Small Is Beautiful (Schumacher), 155
Social policies, 19
Society, 31, 133, 179; innovation and, 180-181; participation in, 177-178; R&D and, 173-174, 182-183; and technology, 47-48, 164, 170. *See also* Culture
Software, 37
Solla Price, Derek de, 71-72
South Korea, 10, 27, 97, 128, 136, 141, 142; economy of, 12, 28, 29; as NIC, 137, 187
Space programs, 39
Specialization, 59
Standard of living, 113
Stanford University, 39
Steam engine, 50, 52, 53, 56
Steel industry, 50
Sweden, 39
Switzerland, 37, 39
Syria, 12, 126
Szent-Györgi, Albert, 73

Taiwan, 12, 28, 29, 67, 97, 128, 141, 142, 187
Tanzania, 30
Tass, 120
Tata Institute, 76
Taxation, 127
Technical Center for Aeronautics (Brazil), 138
Technical change, 53-54, 158, 164
Technical systems, 6, 122; evolution of, 49-51, 53
Technico-scientific system, 56
Technological Prospective for Latin America (TPLA) Project, 177, 180-181
Technology, 2, 26, 33, 36, 133, 138, 165; alternative, 155-156; and capitalism, 134-135; culture and, 21, 172, 196-197; dependency and, 45, 181-182; in developing countries, 68-69; and development, 5, 13-14, 26, 31, 186-187; and employment, 106-107; evolution of, 52-53; hard and soft, 151-152; and industry, 19-20; introduction of, 154-155; in Japan, 14, 74; in Latin America, 42-46; mastery of, 91, 92; and science, 55-56, 158, 189; social and cultural factors in, 18-19, 47-48, 170, 181; Third World, 40-41, 193-194; trade in, 63-64, 86. *See also* Information technologies; Technology transfer
Technology transfer, 63, 64, 128-129, 167, 183; and culture, 133, 152; and education, 21-22; impacts of, 121-122; social process of, 124-126
Telecommunications, 1-2, 37, 39, 53, 91, 97, 117, 119; and computer industry, 93-94
Telephones, 93-94
Television, 94, 119
Thailand, 30
Third World, 3, 104, 159; development strategies in, 156-158; diversity of, 27-30, 185-186; information production in, 120-121; information technologies in, 4-5, 105; new world order and, 179-180; perceptions of, 169-170; R&D strategies in, 181-183; regulations in, 74-75; science and technology in, 40-41, 77-78, 166-167, 193-194. *See also* Developing countries
Tiananmen Square, 144
Tin production, 30
Togo, 156
TPLA. *See* Technological Prospective for Latin America Project
Trade, 11, 25, 27, 118, 136; arms, 127-128; in computer industry, 96-97; of high-tech products, 37-38; in technology, 63-64, 86
Transnational firms. *See* Multinational firms
Transnational organizations, 5-6
Transportation, 39, 111
Turkey, 93

UNCTAD. *See* United Nations Conference on Trade and Development
Underdevelopment, 13, 27, 31-32, 80, 104, 170, 196, 198; and technology, 2-3, 45
Unemployment, 99, 107, 109
United Kingdom, 24, 37, 38, 39, 93, 120; foreign scientists in, 79, 80; Industrial Revolution in, 33, 34
United Nations, 23
United Nations Conference on Trade and Development (UNCTAD), 63
United States, 21, 28, 37, 43, 87, 94, 103, 108, 112, 118, 120, 136; computer industry in, 93, 95; foreign scientists in, 79, 80; and Industrial Revolution, 33, 34; information technologies in, 92, 107; R&D in, 38-39, 40; solar energy project of, 123-124
U.S. Department of Commerce, 36
U.S. Department of Defense, 38, 39
U.S. Department of Energy, 123, 124
United Technology, 39

Universities, 137, 192; R&D at, 39, 71, 72-73, 95-96, 139-140; working conditions at, 75-76
University of Tokyo, 14
Urbanization, 25, 111
Urban renewal, 39

Venezuela, 29, 30, 45, 46, 86, 93
Venice, 35

Warfare, 34-35
Watt, James, 53
Weapons, 34-35, 127-128, 139, 176
Weaver, W., 53
Weber, Max, 17

Wei Jing-Sheng, 144
Weizenbaum, Joseph, 118
Welfare, 48, 126
Wiener, Norbert, 53
World Bank, 13
World War II, 56

Yong-cheng, 132
Yugoslavia, 28, 40

Zahlan, A. B., 76-77, 196
Zaire, 11, 30
Zambia, 93
Zimmerman, Francis, 115, 194
Zweig, Stefan, 138

About the Book and the Authors

This lively book looks at the issues of development in terms that attack both the earlier idealism and the current mood of cynicism about the Third World.

Salomon and Lebeau consider why the great majority of Third World countries have failed to solve the problems of underdevelopment by relying on science and technology, while a very few of them—the newly industrialized countries—have at least partially succeeded.

Opposed to the smug optimism of scientific enthusiasts (though equally opposed to the dismal prophecies of others), the authors argue that, while technological advances may speed the process of modernization in isolated instances, they cannot induce the social transformations that are a prerequisite of development. Scientific research and technological innovation can be effective, they conclude, only where social structures, institutions, and habits have first eliminated the "blocking factors" that are characteristic of traditional societies. It is also essential to recognize that less advanced technologies still have much to contribute to improving productivity and living standards and should not be neglected in the search for solutions.

Jean-Jacques Salomon is professor at the Conservatoire National des Arts et Metiers (CNAM) in Paris, where he is director of the Center for Research in Science, Technology, and Society. In 1963 he founded the Science and Technology Division of the OECD. *André Lebeau,* a physicist by training, is head of the French Weather Bureau and is also professor at CNAM, where he teaches the social and economic aspects of space research. He was deputy director of the French Space Agency until 1975 and of the European Space Agency from 1975 to 1980.